The Texas Water War

Jake Street

Copyright © 2010 Jake Street

All rights reserved.

ISBN: 1461025346
ISBN-13: 9781461025344

DEDICATION

In memory of J. W. "Buck" Buchanan,
one of Texas' original "Water Dogs".

To the Bagwell, Texas rancher, Mickey Anderson, who is always warning, "You can live without love, but not without water…"

And to Roma, who knows, "Water is Resource One."

> When plunder
> becomes a way of life
> for a group of men,
> they create for themselves,
> in the course of time,
> a legal system that authorizes it,
> and a moral code that justifies it."
> **Frédéric Bastiat,** *The Law* (1850)

THE TEXAS WATER WAR
A COMPILATION OF FACTS AND OPINIONS

CONTENTS

Part One:
The War Starts and Where We Are

Part Two:
Water Laws and Those Who Push Them

Part Three:
The Response of Landowner Organizations
Author's Notes and Observations

Part Four:
Water Ownership and Use (various states)
The Texas Groundwater Protection Committee
(Excerpts) Texas Water Code

Part Five:
Conclusion: Water Planners vs Ownership

PREFACE

Consequences:

The philosophy of Consequentialism is referred to as "the ends justify the means". The definition of this phrase is quite succinctly summed up in the understanding that it is quite clearly the concept which has justified the atrocities of Hitler, Stalin, Castro, and the use of force by others pursuing a socialist objective.

The idea that any means is okay, means such as the use of violence against political opponents or lying to voters, is untenable. It is inconsistent with the idea that restrictions on individual freedoms or lying to achieve some perceived common good or world peace in some way serves that end.

There is always some "tension" between Ends and Means.

Means refer always to existing conditions as they are, while the End refers to how things ought to be.

Before we do something based on "for the greater good", the action should demand the following considerations:

What sort of consequences count as good consequences?

Who is the primary beneficiary of the action undertaken?

How are the consequences judged and who judges them?

William Howard Gass, Professor of Philosophy at Purdue and Washington Universities, argues that moral theories such as consequentialism are unable to adequately explain why a morally wrong action is morally wrong. He uses the example of an "obliging stranger" who agrees to be baked in an oven. Gass claims that the rationale that any moral theory might attempt to give for this wrongness (it does not bring about good results), is simply absurd. According to Gass, it is wrong to bake a stranger, however obliging, and nothing more can or need be said about it.

In Texas, we welcome strangers. We will, however, on some occasions, sometimes bake someone, obliging or not, who lives here.

Our doing so, of course, has nothing to do with philosophy or ethics; it's just the consequences of their standing in the way of "how things ought to be" – as seen by government.

1-THE REALLOCATION ERA

Every so often, we hear of a new era. Usually, we are either in it or it has passed us by. Looking back, it seems as if every time period had an era or an age.

There was the Industrial era encompassing the turn of the 19th Century, the Depression era of the late twenties and the thirties, the Hippy Era, The Age of Aquarius, etc. You get the idea.

Today, here in the land of blue sky and broken hearts, if we are not in the middle of a new era or age, we are in a war.

Starting in the late 1970's, a new age or era, what some are calling the Reallocation Era, began dawning in Texas; and, today, we're in the middle of it and property owners are in the middle of a water war.

New water laws which take away the rights of ownership are being forced on property owners, which is a legal taking of their property.

It is an age, as laws are dramatically changed, when existing water rights will be transferred from their historic owners to the State of Texas.

It is the largest raid on private property in the history of the state.

The water taken from the owners will be shuffled from areas that have water to areas that do not have water. The have-nots are winning.

Existing plans call for moving water from one area with a water surplus to another area with water scarcity. It is a state-plan based on taking water from what our politicians see as *"low value use to a high value use"*. It is moving water to those areas with the most political power – the most votes.

This is being done under a coordinated governmental sales-job of *"conservation, protection"* and *"the public benefit"*.

It is control dressed up in more letters.

It is a false perception that the state is *taking* private property to protect and save it from those who would destroy it: The private property owners.

"Life, liberty, and property do not exist because men have made laws. On the contrary, it was the fact that life, liberty, and property existed beforehand that caused men to make laws in the first place." (Frédéric Bastiat, The Law)

So, where does the state, the courts, find the right to take away the historical rights of ownership?

Basically, it is occurring so that the rights to the "conserved" or "saved" or "protected" water can eventually be provided to Texas' big water users. If sold to these users, such sales will be subsidized by the state which, in essence, will force property owners, who once owned the water, to (1) pay for the state taking it and, (2) pay for transferring it to the big water users (adding insult to injury).

Water planners call this *"the public good."* But how can destroying ownership rights be *"the public good"*?

This change in law, in ideas, constitutes a major departure from the way most Texans were accustomed to dealing with private property and their ownership of groundwater.

Their belief that it could not occur has – and is – allowing it to occur.

Reallocation is not a new idea. In fact, what is occurring in Texas water law reflects national trends. The traditional observance of property ownership rights is rapidly disappearing as the nation undertakes more and more reliance on centralized planning, and the implementation of more and more programs for *"the public good"*.

There are more and more of those who tell us that no one really has any rights at all. They are the ones who believe they can grab the rights in their gunsights. They are taking away rights that belong to somebody else.

It usually starts with someone who believes they have the political influence or an ability to *"get what other peoples got..."*

Common sense or a reasonable standard of behavior based on The Golden Rule is not part of their philosophy. But it is, however, knocking on the door of the same basic personal shortcomings exemplified in the history of Stalin, Genghis Khan, Hitler, and others clawed and squeezed to the top in centralized systems.

In society today, too many people believe that the majority of men must be subservient to government. They see government acting to take away some freedom or right from others as something desirable and laudable. It is a confused belief in what is a public good, interest or benefit.

This "Reallocation Era" will be remembered as that period in time when the state's courts turned their backs on historical property ownership rights, giving such rights to the state and their agencies.

It will be remembered as the era when property rights underwent a fundamental change...and the taking of water rights was the largest raid on property in the history of the state.

As is too often the case in politics, it sometimes can become difficult to know just who is sponsoring what or who is speaking for whom; especially when clarity on who, what or the objective is not desired.

The *"texaswatermatters"* organization is sponsored by the Texas Living Waters Project, which is a non-profit endeavor promoting comprehensive water planning in Texas.

In turn, the Texas Living Waters Project is a *collaborative* effort *of the* National Wildlife Federation, Environmental Defense, and the Lone Star Chapter of the Sierra Club.

As politics are played, it seems all these different groups using different names are basically the same, and are just one group.

(Sorry, but generally, the author is highly skeptical of such duplicitous actions, where support for a particular cause can be magnified beyond the reality for the naïve and unwary.)

The fact is, *"comprehensive water planning"* is code for *"control"* of water resources.

Government control has long been the objective of those on the left, which is almost unbelievable, even for a superior race.

Those spinning such words as *"local, conserve, protect, secure, future generations, etc.,"* are urging control. They never seem to consider the fact that when they enable government to take something from others for the public interest, they are enabling government to take something from them when government wants to do so.

They push establishment of groundwater districts as *"local control"* to carry out state rules, regulations, policies, mandates, etc., and pervert the meaning of such terms as *rule of capture* and *ownership*; and generate fear by saying, "the biggest pump pumps the most water" and "protect you against your neighbor", all with the purpose of hiding the true meaning of *"control."*

Control is always government control, either directly or by permission.

Ask your "local" school board.

And it is control aimed at the property of Texas' landowners.

It is also control for the purpose of taking water the state does not own in one area where there are only a few votes and moving it to another area which needs additional water resources and where there are a lot more votes –

And they think that only government can do it.

By luck, the NE Texas area is where the water is……at least for now.

But the water-planners, lawmakers, environmentalists, social scientists, law-firms that benefit from legal battles, and those who most always believe control is a good thing for society, are looking thirstily in this large corner of Northeast Texas.

And these groups are counting on local landowners muddying up the water by doing nothing. Only one NE Texas group, Rural/Urban Resources, is fighting to keep reason and the rule of law working for the property owner.

Many of Texas' larger organizations representing landowners, have signed off on state control of all groundwater. They have done so because they are often managed by individuals who also act as lobbyists. Thereby, these managers are receptive to the arguments of their counterparts they see and deal with everyday. Many are on friendly terms with the water planners.

Failing to see or understand the danger to ownership rights, they offer advice and input back to the organization that place in danger the very landowners these managers are paid to protect.

Being blind is not limited to those without sight.

The water planners know where the water is, and when they're ready for it, they'll go and get it.

2-ODDS AND ENDS

Direct one-piece legislation, stating the absolute control of groundwater rights, as such, will likely never be enacted by the Legislature. Lawmakers, state as well as national, have ways of accomplishing things without there ever being an open direct vote on an issue. But all political actions and trends in water politics are headed toward *state-control of water resources.*

Control has come (and continues to expand) a little bit and a little bit at a time, incrementally. Policies, guidelines and recommendations have been formulated piecemeal, slowly approved and accepted as "the state's policy" or as an "agency desired way" of doing a particular thing. Enough legislation has been enacted to establish the process and the intent – and to change the meaning of ownership.

Major water legislation was enacted by "merely ratifying" or "giving formal consent to a long-standing state policy…"

Piecemeal formulated policies, guidelines and legislative acts, slowly approved and accepted as "the state's policy" or as an "agency desired way" of doing a particular thing, were the process and the intent, whereby the water planners who sought control of underground water took control of the landowner's property.

These shenanigans were designed so that an unpopular vote to actually take control of underground water rights will never be necessary.

It gives politicians a way to enact something without ever actually voting on it...

This, in effect, allows a policy granted to a state agency to become "a legal fact" —

And it doesn't require a legislator's vote to have a state-wide Master Plan (for the development of Texas' water resources), which includes taking control of a landowner's property.

It never requires giving voters a state-wide Master Plan for their approval.

And the political fallout will be less.

Those pushing districts depend upon the stupidity of those who see no evil in taking a landowner's property.

Legislators have also given groundwater districts the power to develop their own "water fields", drill their own wells, and to transport the water from these "fields" to a market over long distances.

Technically, Texas' lawmakers and their supporters, like too many millions of our nation's citizens, claim that control is not confiscation. But that is a distinction without a difference.

If you control it, does it matter who owns it?

Isn't confiscation plunder?

Legislators have given millions of taxpayers' dollars to those pushing for taking or controlling groundwater.

The tax dollars of Texas property owners are being used to publicize, promote and sell the taking of their property.

That, in itself, is evil – and there is no other distinction or difference.

For years, property owners have elected legislators who believe that "nothing will ever get done" if citizens are allowed to vote on a program or matter that requires imposing a tax or new taxes on themselves. Basically, believing that voters are too dumb to know what is best.

Ironically, these same legislators believe those same "dumb" voters are smart enough to elect them to office.

Or maybe, dumb enough to elect them to office.

3 – WATER AS PROPERTY

Today, Texas property owners are possibly facing the greatest takeover of private property in state history.

For over a century, the state's property owners have owned the groundwater beneath the land surface of their property.

This historical right is now under attack.

The last legislative session in Texas resulted in 26-different bills concerning underground water being passed into law – and not one of them came close to protecting the historical ownership of the water beneath the surface of a landowner's property. Not one.

But each of the 26-bills addressed some form of control of the landowner's property.

The Texas War over water started in the mid-1960's, when Water, Inc., headquartered in Lubbock, was organized to import water from the Arkansas and Mississippi Rivers to West Texas, Eastern New Mexico, and the Oklahoma panhandle.

The water was to be pumped from the Arkansas River, up the Canadian River, to Lake Meredith, near Amarillo, Texas. The lake would be used as a temporary storage location for the imported water. From Lake Meredith, the water would flow by gravity thru' pipe and canals to the southern High Plains area of Texas and eastern New Mexico; and be pumped into the northern areas and to the Oklahoma panhandle. Surplus water would be injected into the underground aquifer for long-term storage.

And here came reality: Once the imported water was injected into the aquifer, it would mix with and be unidentifiable from the existing water in the aquifer.

By 1967, as it was realized that imported water injected into the aquifer could not be identified, it created three questions:

"What was to stop a landowner from pumping this underground water for his own irrigation purposes?"

"How could you measure how much imported water a landowner was using? And

"How could you demand a landowner pay a 'fair share' of the imported water's cost?"

The Water, Inc., effort failed, primarily because of internal politics, originating over control of the organization by some agricultural and business leaders in the South Plains area of Texas.

But these 1967-hypothetical questions started a discussion about groundwater in the state's growing professional water management circles. They were the starting point for later efforts seeking control of all groundwater in Texas.

For the last half-century, water planners all over the nation have been fighting to re-define ownership of water for the sole purpose of controlling it. They have succeeded in establishing inroads on groundwater as a private property in many states.

As in Texas, the planners have done so by making administrative definitions that have been accepted by a careless public and lax courtroom authorities.

As is now being done in Texas, planners have successfully argued that legal classes of water were increasingly subject to poorly-defined common law water rights and, therefore, needed strict administrative definitions and statutory regulation. This is an argument that you can make about most anything that another person has – and you want.

In several states, the strict common law of absolute ownership was legally forced to accept two variations, the doctrines of reasonable use and correlative rights. It is these two less protective ownership theories which are being pushed in Texas today.

It is this pattern, however, as history shows, by which freedoms are lost: Define the definitions, the meanings of words, get the lawmakers to accept them, and the courts will follow.

From the author's perspective, definitions are important. And because they are, we need to have a fairly common understanding of definitions of property and ownership, which can be found in the pages of *Wikipedia*, the free encyclopedia:

First, property is any physical or intangible entity that is owned by a person or jointly by a group of people. Depending on the nature of the property, an owner of property has the right to consume, sell, rent, mortgage, transfer, exchange or destroy their property, and/or to exclude others from doing these things.

Important widely recognized types of property include:
- Real Property (land),
- Personal Property (physical possessions belonging to a person),
- Private Property (property owned by legal persons or business entities),
- Public Property (state owned or publicly owned and available possessions), and
- Intellectual Property (exclusive rights over artistic creations, inventions, etc.), although the latter is not always as widely recognized or enforced.

A title, or a right of ownership, establishes the relation between the property and other persons, assuring the owner the right to dispose of the property as he sees fit.

There is a certain amount of disambiguation concerning property rights. The authors of the American Constitution, as well as other philosophers, found and find origins of property rights in morality or natural law. Most advocates of a strong centralized structure of society believe, however, that all property rights arise from social convention.

It is the growing distinctions between the legal plunder of property and the moral right of property ownership that are creating opportunities for those who seek control of society.

In Texas, the advocates of centralized control are winning.

Historically, the American Constitution established property rights under absolute ownership, a doctrine honored in a history of law.

Under absolute ownership, Texas landowners own the water beneath their property, but this may not be the case much longer.

Property owners, who voted to give control of their groundwater to a water district, are now sweating the forthcoming decisions of a case (*Edwards Aquifer Authority v. Day, 274 S.W.3d. 742 (Tex. App. – San Antonio 2008, pet. pending)* that will determine whether control of groundwater is a taking of vested private property and, if so, does it require compensation?

Briefs have been filed by the Edwards Aquifer Authority, the Texas Alliance of Groundwater Districts, and the Texas Attorney General claiming that there is no ownership interest until water is actually produced and reduced to possession.

Under this theory, groundwater districts and the state could unreasonably regulate groundwater – in fact, completely halt its production – or use it themselves - without any liability for depriving a property owner of that property right.

If adopted, this position would completely undermine the private property rights of all Texas property owners. It would mean that the state could take your property and pay you nothing for it!

4 – PROPERTY RIGHTS

Now is a good time to look at property rights – to more fully understand what the planners are planning to circumvent.

The first property right an individual has is to his own body. How can you not have a right to your own body? And as you have that right, it also follows that you also have the right of personal liberty. These are the first principles in the theory of absolute ownership.

The principle is simple: There is no freedom without property rights.

As the song states it, *"You can't have one without the other…"*

Another way to state it is there are no "human rights" which are not property rights.

Today, the concept of "human rights" has evolved to demand that an obligation be placed upon others to provide whatever it is that someone believes he has a "right" to: healthy food, decent housing, medical care, education, transportation, etc.

They are demanding "rights" to property. The result of any such demand is that someone has less of something, even if it is only money.

But centralized planners have given us the idea that "human rights" are separate from property rights.

How do you separate the two?

Generally, "human rights" are those that weaken all rights on behalf of the "public benefit" or the "public good." Thus, they claim that "human rights" take precedent over "property rights."

Centralized planners cannot operate well when property rights are absolute, as they need fuzzy and unclear standards by which to achieve their goals.

As we face a growing centralized society, the planners often refer to "free speech" as being part of "human rights", but forget that both are grounded in private property rights.

The U.S. Supreme Court Justice Hugo Black, a well-known "freedom of speech" advocate, made it clear when he stated: *"I went to a theater last night with you. I have an idea if you and I had gotten up and marched around that theater, whether we said anything or not, we would have been arrested. Nobody has ever said that the First Amendment gives people a right to go anywhere in the world they want to go or say anything in the world they want to say. Buying the theater tickets did not buy the opportunity to make a speech there. We have a system of property in this country which is also protected by the Constitution. We have a system of property, which means that a man does not have a right to do anything he wants anywhere he wants to do it. For instance, I would feel a little badly if somebody were to try to come into my house and tell me that he had a constitutional right to come in there because he wanted to make a speech against the Supreme Court. I realize the freedom of people to make a speech against the Supreme Court, but I do not want him to make it in my house."*

The authors of the American Constitution sought to establish an absolute ownership of property, balanced with a rationale that government would be restricted from arbitrary power over the property owner and their estate. In essence, limiting what a government might or might not do.

Isn't it ridiculous to think that men, who had personally faced and fought against the absolute arbitrary power of the King of England, would give to any form of government any of that same absolute arbitrary power over their person or property?

After risking their lives, why would they put themselves back into the boiling pot they had just (barely) escaped?

For the first time, a new country had the power to execute the terms of a liberty, the freedom they had struggled against all odds to achieve.

Who in their right mind could conceive these individuals would voluntarily place themselves under the absolute power of another – especially, a legal system or a group of legislators, or judges?

For the first time in history, they had the power to defend themselves and their property against arbitrary power, and the power to protect against those who would make a prey of them.

And they did.

In the 17th Century, John Locke, wrote that both persons and estates are to be protected from the arbitrary power of any magistrate, including the *"power and will of a legislator"*. He argued that depredations against property were a plausible justification for resistance and revolution just as those against persons. In neither case, Locke believed, were subjects required to allow themselves to become prey.

The authors of the Constitution were familiar not only with the writings of Locke, but also those of 18th Century philosopher William Blackstone, who explained the ownership of property.

In the 1760s, Blackstone, in his *"Commentaries on the Laws of England"* wrote that *"every wanton and causeless restraint of the will of the subject, whether produced by a monarch, a nobility, or a popular assembly is a degree of tyranny."*

Blackstone argued that only through property rights could such tyranny be prevented or resisted. That was why he emphasized that indemnification (payment in value) must be awarded a non-consenting owner whose property is taken by eminent domain. Blackstone also said that a property owner is protected against physical invasion of his property by the laws of trespass and nuisance.

One of America's earliest U. S. Supreme Court Judges was James Wilson, previously a professor of law at the University of Pennsylvania. In the early 1790's, Justice Wilson wrote and presented a series of lectures on the *History of Property*, in which he openly declared: *"Property is the right or lawful power, which a person has to a thing."*

He presented two philosophical premises: *"Every crime includes an injury: Every injury includes a violation of a right."* And he focused on the question of whether man exists for the sake of government, or does government exist for the sake of man – the question of natural and absolute rights and whether property is one of them. Judge Wilson wrote: *"In his unrelated state, man has a natural right to his property, to his character, to liberty, and to safety."*

In his essay, *"The Natural Rights of Individuals"*, Wilson asked if the institution of government was to acquire new rights for government, or was government to acquire security for the possession or the recovery of rights.

Wilson then presents the conclusion that exclusive property ownership, as opposed to communal property, is to be preferred.

In Texas today, too many of our legislators, too many planners, too many voters, have failed to appreciate or understand what we are doing to the principles of property ownership.

And now we are destroying our individual freedom.

5 – ABSOLUTE PROPERTY OWNERSHIP VS VESTED RIGHT

The case *(Edwards Aquifer Authority v. Day, 274 S.W.3d. 742 {Tex. App. – San Antonio 2008, pet. Pending})* that will determine whether control of groundwater is a taking of vested private property and, if so, does it require compensation, is a big deal.

It has the power to drive a stake into the heart of property ownership.

Basically, the court is debating an effort to change the historical absolute ownership of property to a vested right.

The legal beagles are barking up a storm about how a vested right is one that is in addition to absolute property rights. But the only reasonable conclusion that can be drawn from consideration of the two is that one is secondary to the other: Vested is not absolute.

If property owners have absolute rights, then why are any secondary rights needed?

But wars aren't fought for truth and justice.

The big danger to property owners is that by hearing the case *(Edwards Aquifer Authority v. Day)*, regardless of what it may decide, the court has opened a door which limits absolute ownership: The court basically is debating an effort to change the historical absolute ownership of property to a vested right.

Courthouses are full of attorneys with differences of opinions formed by opened wallets.

But basically, a vested right is part of a property protection doctrine that is not necessarily grounded in constitutional law protecting absolute ownership, but rather notions of fairness and equity (personalty). It is a way of looking at property that establishes "a part of ownership recognized or given" –

For instance, if you leave a painting to a niece or nephew in your will, they would have a vested right to the painting, but not to any other of your property.

Another way of looking at the difference is at employees who have a vested right in their company's retirement plan: Each has a right to a part of the plan's benefits, but none have the absolute ownership of the plan.

Vested rights are those that are so settled in a person that they are not subject to defeat or cancellation by the act of any other private person. Vested rights have also acted as a limit on the police power found within government itself. In short, it is only a part of absolute property rights.

Vested rights grew out of government taking the property, primarily land, of individuals. Questionable judgmental calls by legal authorities established vested property rights as a way to compromise the evil of taking a property with a payment to the owner – whether the owner wanted to give up the property or not.

It was a way to "take" private property while presenting a public face of fairness. It was a way for government "to get around" absolute ownership.

A "vested right" has come to mean that a governmental unit gives back to a landowner part of the property which it has "taken", to promote some perceived public benefit.

Today, landowners have allowed water control planners and their attorneys to bring all Texans to a point where a court giving recognition to a part of ownership rights is more important than constitutionally-protected absolute property rights.

The Absolute Dominion Rule, absolute ownership, is the law in Texas, and in Connecticut, Indiana, Louisiana, Massachusetts, Mississippi, Rhode Island, and Maine.

Twenty-one states have adopted or indicated a preference for the Reasonable Use rule: Alabama, Arizona, Arkansas, Delaware, Florida, Georgia, Illinois, Kentucky, Maryland, Missouri, Nebraska, New Hampshire, New York, North Carolina, Oklahoma, Pennsylvania, South Carolina, Tennessee, Virginia, West Virginia and Wyoming.

The Reasonable Use Rule is essentially <u>the rule of absolute ownership</u> with exceptions for wasteful and off-site use.

Regardless of what water planners claim, Texas is not the only state that has historically observed the Absolute Ownership rule.

Regardless of what other states may or may not do, Texas has the right to exercise our own State Sovereignty to decide what is best for Texas.

Texas got here by listening to the same history tellers of tall tales who claim they know how to do things better than anyone else.

And the ego-driven conceit of those who are always looking for loopholes to escape moral imperatives.

6 – THE MEANING OF WORDS (1)

(Much of the following has been borrowed from "Going Postal", a novel of fiction by Terry Pratchett; a most delightful book that can be found (hopefully) at your local bookstore or in your community's library):

Whom do you believe?
Why?
Based on What?

Words, describing and displaying love, friendship, anger, hate, terror, exasperation; words appropriate and inappropriate; elegant words and words that buzz and words that whisper and words that shout; big and small words; words of hope and disappointment; words and more words and more words...

All the words in the world are garbage when cooked by a master.

Innocent words are mugged, ravaged, ravished, stripped of all true meaning and decency, and then sent to walk the gutter for the slick and smooth word masters.

A word master might ask you, for instance, what is the meaning of "is"? And some valueless soul applauds the evasion and the lie as something great and wonderful...

Words are not some mysterious spasm in the universe. We hear words and they are recognized as a symbol of some meaning by our brain. Words have nothing to do with greed, arrogance, and willful stupidity, unless spun by a master.

The word masters don't make mistakes, just "well-intentioned judgments" which, with the benefit of hindsight, might regrettably have been, in some respects, in error, but these had mostly occurred, it appeared, while correcting "fundamental systemic errors" committed by the previous management or administration. No one is sorry for anything, because no living creature had done anything wrong; bad things had happened by spontaneous generation in some weird otherworld, and "were to be deeply regretted."

Sometimes, it's all about "the people." As in, "the people want this"...and "the people want that"...and "the people demand it" — Whenever you hear "the people" it's really about the goal of some charlatan promoting the same cronyism, same piggy ways, and same stupid arrogance. *It is one more leech in a chain of leeches making an impressive beacon to oppression and division.*

You hear the words and you know, somehow deep inside, even if you blindly won't or can't bring yourself to admit it, that what you're hearing are the squeaks of a weasel in a tight corner.

Meaningless, stupid words, come from people without wisdom or intelligence or any skill beyond the currency of expression. Words that stand for everything, from life and liberty to Mom's Homemade Apple Pie, can become words that stand for everything, but anything. And nothing.

Charlatans destroying meanings into obscenities.

So, Whom do property owners believe?

Why do property owners believe it?

And based on What?

7 - THE MEANING OF WORDS (2)

For decades, landowners have allowed those seeking control to shape the argument, establish the terms of the debate, and supply the definitions of the words used in the argument.

In a war of words and ideas, if you can control the meaning of the words, the other side is without ammunition.

Sadly, landowners lost the first battle of public perception to the water planners: Most Texans have swallowed the nonsense of the new meaning of the "rule of capture" –

Critics of the "rule of capture" claim it is harsh and outmoded, as it provides no protection for a landowner, saying that "a landowner may pump as much groundwater as he chooses without liability to neighbors for drying up their wells."

They claim the "rule of capture" means "the biggest pump pumps the most water."

They claim that control, doing away with the "rule of capture", is needed "to protect you against your neighbor..." -

All these are statements designed to create fear, as none of them are factual.

Not only are fear tactics used, but property owners are fed misinformation similar to "Texas is the only state in the country that has not yet fully abandoned the rule of capture."(*An Overview of Water In Texas,* Cynthia Cox Payne, October/November 2010 <u>Landman Magazine</u>)

Well, the fact is, as previously stated, the landowner does owns the water beneath the surface of his property in Texas, as well as in the states of Connecticut, Maine, Indiana, Louisiana, Massachusetts, Mississippi, and Rhode Island. These states have historically used the "rule of capture" to protect the property rights of their respective landowners.

In Texas, those pushing control of underground water refer to the absolute dominion rule generally contemptuously, for political reasons, as "the rule of capture".

Several of the states have reaffirmed that the 'rule of capture' "has done no harm".

In 1999, Maine's Supreme Court ruled on an effort by the state to eliminate the "absolute dominion" rule, saying:

"We decline to abandon the absolute dominion rule. First, we are not convinced that the absolute dominion rule is the wrong rule for Maine…Although modern science has enlightened our knowledge of groundwater, this does not mean that the rule itself has interfered with water use or has caused the development of unwise water policy…Furthermore, for over a century landowners in Maine have relied on the absolute dominion rule. In the absence of reliable

information that the absolute dominion rule is counterproductive and a hindrance to achieving justice, we will not depart from our prior decision."

Please note the above: "...*does not mean that the rule itself has* interfered *with water use or has caused the* development *of unwise water policy...*"

Unfortunately, most landowners and voters don't know that the "rule of capture" - which those seeking control frown on - is older than Texas.

The Magna Carta (1215) was arguably the most significant early influence on the process that led to the rule of constitutional law. It influenced our own United States Constitution. It was the first document forced onto an English

King to attempt to limit the King's powers and protect the privileges of his subjects.

By the late 1200's, clauses had been added to the Magna Carta stating that no royal officer may take any commodity such as grain, wood or transport it without payment or consent...

In America, the first ruling on underground water was related to this older English Common Law; establishing that if a deer, a fox, a duck, a rabbit, (whatever), was shot by someone else, somewhere else, but *dies* on a landowner's property, it belonged to the landowner.

It is this basic historic principle that is referred to as *'the rule of capture'* - and its historical intent has been to *protect a property owner* against another taking possession of the landowner's property.

This was the principle used in the 1904 Supreme Court decision (Houston & Texas Central Railroad Co. v. East, 81 S.W. 279), which protected a landowner's use of groundwater.

But property owners have allowed a debate change from protection to one of control.

Landowners should be arguing that *IF* "a right" to property under absolute ownership exists that, as night follows day, "a right" to what is *on* the landowner's property exists and that, as day follows night, there is "a right" to what is *under* the owner's property.

IF *one* of these rights does not exist under the law of absolute ownership, then future planners will have an argument that none of these rights exist and are subject to their control.

Any intent to do away with the so-called "rule of capture" - should strike fear in every property owner, especially mineral owners.

But landowners have allowed the water planners seeking control to establish the perversion that doing away with the intent of the "rule of capture" (to protect the property owner) is a good thing...

Those creating fear between neighbors need to understand a more likely scenario: The owners of the "hundreds of acres" around the big pump – if they want it – are the *first* to get a lion's share of any water *moving toward* the "biggest pump".

It's not the size of the pump that counts, but the amount of the water under it! And not one aquifer in America has ever "gone dry".

Fear can't change facts.

Underground, water moves very slowly. The aquifer controls how fast water moves, not a pump or an underground water district. Only Mother Nature controls that…to claim otherwise is wrong…to think otherwise is arrogance.

Who is protecting property owners against control?

Who is protecting property owners against government taking their property?

If property owners do not take the time or make the effort to shape or influence the meaning of the words used, of arguments being used for control, they will lose.

8 – SELLING GROUNDWATER DISTRICTS

It is time, as the Walrus said, "…to speak of many things." Especially, of the many things about groundwater districts.

Unlike what they want the public to believe, groundwater districts can be a big pain in the property owners' *prosperity*.

When property owners give control of their underground water to a water district, the district takes it and never gives it back.

They don't have to, as legislators have written into the Texas Water Code a provision that prohibits a water control district from ever being disbanded or dissolved.

Once a district is established, it is there for eternity.

We are told by the Texas Water Development Board that approximately 85% of Texas' land surface is in a water district of some kind. But in January, 2011, the Texas Groundwater Protection Committee, in its report to the Texas Legislature, said that about "100 counties" were not in a control district.

But regardless of land surface or the number of counties in a district, property owners were and are told that "a water district is essential to protect your water rights".

But few of those voting for a district ever asked themselves, "When did my ownership of the water beneath my property change to mean that I only have 'a right' to it?"

A "right" is something that can be given or taken away by those who control it.

Property owners never asked "Why would those who are telling me they will give me 'a right' to water I own, protect it? And how does my giving them control of my water protect me?"

If you control it, does it matter who owns it? Isn't control basically having the power to seize the landowner's asset and take it for the use of the one who controls it?

We all pay taxes which help build highways, but we don't have a right to drive on them: The state forces you to pay a fee for a driver's license, and the state claims it gives you "the privilege" of driving on the highways.

This is how a water district works: You pay a fee for a permit and the district gives you the privilege of withdrawing so many gallons of your water, or it may prohibit any withdrawal of a property owner's water.

Their permission is the final word.

We are told that a district is a water "conservation" district.

Well, who are they conserving the water for? It certainly isn't the local landowners.

Only recently have water districts started emphasizing the "conservation" in their names, primarily for reasons of perception (politics): "Conservation" sounds much better than "control" –

The ones seeking control talk about a district "protecting you from others pumping too much water," or "protecting you from your neighbor" - hoping property owners don't realize this also means "protecting your neighbor from you".

The last question those pushing control want property owners asking is "Who will protect me and my neighbor from the 'conservation' district?"

Property owners heard the claim that metering was/is imperative, as monitoring use is in the landowner's best interest, as it protects the landowner from others pumping too much water from the aquifer. (Metering is mandatory for all wells, existing as well as new.)

Property owners never substituted "control" for "monitoring" and "metering" –

They never stopped to ask "*Why* is it in my best interest?"

The water planners never want property owners asking, "*Who* will set 'allowable' pumping amounts; how much is 'too much'; what will a permit cost; *who* decides if I will get a permit; what will a meter cost; and how much is this district going to cost me?"

Those selling the districts made ridiculous claims that a district was the only way to have "local control" of water resources.

But the state decided that local governments in local counties cannot or could not provide the type of groundwater regulation the state wanted, so know-nothing legislators and water planners gave state agencies the power to force districts on local areas.

There's nothing local about state laws and regulations *forcing* "local areas" into districts! As landowners owned the water beneath their property, wasn't it already under local control?

What most individuals too often fail to remember is that state agencies are organized to control local areas. For instance, look at how much time is necessary for local schools and county commissioners to waste in dealing with state regulations and mandates.

Being forced to do things by some state mandate isn't local control . . .

When the state determines the operating policies, regulations, goals and objectives for all water districts, and a district's headquarters can be hundreds of miles away, there's nothing "local" about it.

When government forces you to pay them to take control of your assets, government is not your friend –

9 – GROUNDWATER DISTRICTS

Everywhere, it seems property ownership rights are under assault from various powerful groups who not only approve, but actually urge, government control of private property.

Many millions of supposedly sane citizens see no harm in giving control of property to government. They never seem to consider that control equals or surpasses ownership. As we continually ask, "If you control it, does it matter who owns it?"

The United States Constitution says that private property is not to be confiscated by the government without due process of law. But lawmakers have created laws which weaken or circumvent the rights of ownership.

Too many of us were and are willing to let the law become laws of men.

In Texas, where groundwater has been historically owned by the owner of the surface acres, lawmakers have given the power to control private property to groundwater districts.

Not only did a majority of the state's legislators give control to these districts, but they also gave the power of taxation.

By law, a district must have a minimum annual budget of $250,000 from district property taxes or from a combination of district taxes and fees.

In essence, the control takes away the rights of ownership, and the taxation forces the landowner to pay for the control of his property by others.

You must give the Devil his due: It took an evil mind to think this scenario up . . .

Legislators have also given the power to create such districts to units of state government, without a vote of the people who live in the area of the proposed district.

The point being that the last thing those pushing a control district want is for landowners to vote on it.

The irony missed or ignored by those who demand a water district without a vote, is that the very people they're asking can only achieve office by a popular vote.

We should never elect to public office anyone who would even consider circumventing a right to vote. To take away, to deny, a right to vote on matters that affect individual lives, incomes and property is reprehensible. Contemptible.

Neither can we trust anyone who would advocate doing away with our right to vote.

But a majority of Texas legislators, evidently, had no qualms about doing so.

Districts now are created by the legislature *or by the Texas Commission on Environmental Quality* with the sole purpose of controlling groundwater. Such districts can be created by one of three procedures:

(1) Special law districts can be established by the legislature (introduction and passage of a bill to do so);

(2) Districts can be created in priority groundwater management areas through procedures initiated by the TCEQ (Sections 35.012(b) and 36.0151 TWC). No vote required. And,

(3) Districts can be created through a property-owner petition filed with the TCEQ (Section 36.013, Texas Water Code). Generally, this is by a majority vote of those living within the boundaries of the proposed district.

Districts are local or regional in their jurisdiction.

Those pushing control of underground water act and think as if they were wearing halos. What they fail to realize is that if the halo were to fall, it would be a noose.

Most of them are long on promises, but short on memory.

Those pushing control are full of "protect our water" and "protect our future" comments. They keep forgetting, it's not their water!

Underground water belongs to the landowner. So, how does it get to be "our water"?

And, please, what is this "protect our future" nonsense? Just who is this "our" they're protecting?

And they're protecting "our future" by taking control of the landowner's property?

Again, if you control it, does it matter who owns it?

Maybe, since water districts can establish their own water fields, take the water, pay landowners nothing, and sell it to their friends, the Land of Oz, or whomever they like, a water district is the preferred method of those who will control the water district?

Somewhere along the line, property owners should have understood that districts could take their water and sell it to others, without paying the landowner one dime for it.

For instance, (1) in 2003, the Texas General Land Office created a water-storm with its negotiation, through the School Land Board, to lease with private entities the rights to pump water from public land. And (2) the Aquifer Group, a private-public water marketing firm, has met with members of the Legislature concerning their plan to commercially sell groundwater from state lands that lie within a West Texas Groundwater District.

As water planners were screaming about "conservation" and "local control" and pointing with fear at "your neighbor", the state was doing what the state's groundwater district has the power to do!

While water planners have created an enormous amount of fear by screaming "the biggest pump pumps the most water", districts actually have little to do with "the biggest pump pumps the most water":

For instance, the South Plains Lamesa Railroad v. High Plains Underground Water Conservation District (No. 1, 52

S.W.3d 770, 2001) case provides an example of a district acting in a fashion that a court found to be unreasonable.

After a permit had been paid for and given, a well was drilled and equipped at a cost of $30,000. After completion, the district passed a motion revoking the permit, citing existing deficiencies. When the applicant re-filed an application that remedied the alleged deficiencies, the district denied the new application "to prevent a disproportionate taking of water."

The district was sued. Their action in revoking and denying a permit were found to be improper by the High Plains court, because the district's rules contained no provisions that would authorize denial or revocation of a permit because a well would produce a disproportionate amount of water.

In addition, the court held that the action of the District prohibiting "a disproportionate amount of water to be pumped as it relates to tract size" was not otherwise authorized by statute because (1) such authority was not clearly authorized by the Legislature, (2) the statute did not provide reasonable standards to guide the District in exercising its powers, and (3) the District was not authorized to deny a permit to prohibit a disproportionate amount of water to be pumped as it relates to tract size based upon its alleged discretionary power.

As you might imagine, steps are being taken by the Texas Water Development Board and districts statewide to remedy this lack of control.

Since 1983, almost every lawsuit concerning water and settled in a Courtroom, originated in an underground water district. Yet, supporters claim districts "protect you against your neighbor..."

When water problems are originating within districts, primarily against districts, it appears that "you and your neighbor" will be getting more hassle than protection.

Alzheimer's, arrogance, moral degeneracy, and other such symptoms, seem to be the prevailing requisites for creating a desire to control the assets of others.

What's next, passing a bunch of regulations forcing victims to pay the ones committing the crimes? Why not? Isn't it what is actually being done, behind the dancing of legal money-raking, to the landowner?

There are laws against changing the brand on another man's cows and hauling them off to market. It's called "rustling". But a water district can put its brand on the landowner's water and haul it off to market.

What is the moral difference?

10 – A SHORT HISTORY OF TEXAS WATER

The Texas Water Development Board has instructed Regional Water Planning Districts, Groundwater Management Areas and the state's water districts to claim they know how much water is in an aquifer, and that up to 90% of it is under their control. (The plan is based on assumptions and estimates, which are suspect.)

Different districts are making claims that the water in aquifers in their areas, in amounts ranging from 40-to-90-percent of the groundwater, belongs to the district.

In areas where there is not a district, Regional Water Planners and Groundwater Management Areas are assuming the power to act as one.

More lawsuits are coming.

But how did we get here?

As mentioned earlier, Texas' groundwater law is based on the English common law rule of "absolute ownership." It means landowners have the right to capture the groundwater beneath their land because they own it.

It was adopted by the Texas Supreme Court in 1904 in Houston & T.C. Ry. Co. v East. The Court determined that one has a right to dig a well and capture all the water in it, and that no correlative rights with regard to underground percolating waters were recognized by law. This ruling became known as the "rule of capture".

In justifying its decision, the Court said, in part: "Because the existence, origin, movement and course of such waters, and the causes which govern and direct their movements, are so secret, occult and concealed, that an attempt to administer any set of legal rules in respect to them would be involved in hopeless uncertainty, and would be, therefore, practically impossible." For years, Texas' courts treated the rule of capture as a tort doctrine.

Years later, in Texas Co. v. Burkett, the Texas Supreme Court fortified the rule of capture by interpreting it to be a vested property right rather than a tort doctrine. In Burkett, the Court ruled that the plaintiff "plainly had the right to grant access to and the use of their waters for any purpose...."

This created a common law rule that also classified water ownership as a vested property right.

In 1949, and again in 1985, the Texas Legislature passed legislation acknowledging private ownership of groundwater, while allowing the formation of water districts.

Acknowledging the common law doctrine, the Texas Legislature codified the absolute ownership principle when they enacted the *Texas Underground Water Conservation Act of 1949*, which recognized ownership and rights in underground water of "the owner of land, his lessees and assigns."

Two years later, in 1951, the first Groundwater Water District was created with headquarters in Lubbock, Texas.

Then beginning in the 1960s, state water planners began to pressure the Legislature to control underground water.

In 1977, less than 30-years after the 1949 Water Act, in 1977, the seventy-fifth regular legislative session passed Senate Bill 1, which expressed the belief that "water districts are the best way of countering the 'rule of capture'…controlling underground water…"

Senate Bill 1 was the first overt effort to circumvent the absolute ownership of groundwater as recognized in Texas

Supreme Court rulings, the Texas Water Code, and as expressed in the 1949 Water Act.

SB 1 was a compromise product of Lieutenant Governor Bob Bullock, and its authors (Senator J.E. 'Buster' Brown and House sponsor Representative Ron Lewis), and numerous state agencies that were seeking more power and more control of the state's water resources.

Since the late 1970s, the objective has been for the state to assume control of groundwater.

In 2001, at the urging of the Texas Water Development Board, the 77th Legislature passed Senate Bill 2, which authorized the TWDB to create sixteen (16) Regional Water Planning Groups and 16 regional groundwater management areas (GMAs). Publicly, the reasoning behind the measure was to *assist the TWDB in long-range planning for the development of Texas' water resources.*

In actuality, under the guidance of the TWDB, these regional groups were to act as the "planners" and "managers" of water resources in their areas. This would provide political cover for the state water planners. Through its design, the "planners" and managers hoped that these new creations would be seen as "experts" recommending what needed to be done for the development of Texas water resources "for the public benefit."

The Groundwater Management Areas were to eventually become 16 "super" groundwater districts, with "existing" districts carrying out their directives.

These goals were achieved in 2005, when the legislature adopted House Bill 1763, requiring the regional water planning groups to use the groundwater availability numbers developed by the *GMA's.*

This gave the GMAs authority for joint planning within the boundaries of a groundwater aquifer. And the bill included provisions for GMAs to act as water districts (with control of groundwater). This meant that existing districts would become "offices" for these more-powerful regional districts.

Today, many of the governing boards of the state's groundwater districts are unaware of the state's plan to take control of the districts they serve.

Senate Bill 2 outlined how these districts were to be replaced with the 16 Groundwater Management Areas.

As government planners know, it is easier to control a few that control many.

Currently there are about 95 groundwater districts in all or part of about 150 counties. Just under 100-counties remain free of a district's control.

Many of these counties don't want the control that comes with a district, and several have rejected efforts to establish a district.

The problem facing the planners was how to take control of groundwater where there was not a district to act at the state's direction.

Placing all of the state in 16-Groundwater Management Areas, and giving these GMAs the powers of a Super Control District, means "local districts" are no longer needed to take control of all of Texas' groundwater.

Facing Facts: When the experts talk about a Master Plan for "developing the state's water resources" they can only be talking about underground water: The state already controls surface water — they can do what they want with it now!

In 1999, some 95 years after the East decision, the Texas Supreme court reviewed the rule in Sipriano v. Great Spring Waters of America, et al. (1 SW3d 75 [Ozarka].

This case involved a claim by a domestic well owner that Ozarka's (Great Springs) nearby pumping had dried up his well.

The court unanimously affirmed the rule of capture.

The Court's ruling, however, did suggest that it might change the rule in the future, but would wait to see if the Texas Legislature would address groundwater questions adequately.

At this time, the Court is back preparing another ruling *(Edwards Aquifer vs Day.)* A ruling not on ownership, but on "vested rights" and/or the state's assumed right to take property without payment.

Shouldn't the court first decide if the landowner owns the groundwater? Are over 100-years of history, the Texas Water Code, and previous Court rulings wrong?

If the landowner owns the groundwater, and the Texas Water Code states that the landowner does own the groundwater, why is the Court hearing a case that will "give" a vested interest to a landowner?

If the landowner owns the water, shouldn't the Court require that ownership rights be recognized and enforced?

If the property owner does not own the groundwater, the Court needs to (1) clearly state the state owns it, and (2) that the landowner does not have a claim on the water or any part of it.

Honesty is fairly simple.

If we have laws that are unenforceable or laws officials will not enforce, then we are living in the wild and anything goes.

11 – ESTIMATING GROUNDWATER SUPPLIES

(Estimates of groundwater availability have been developed for and accepted by the *Texas Water Development Board [TWDB]*. These numerous local and regional aquifer studies sponsored and paid for by the TWDB, employed various methods for estimating water availability):

Uncertainty causes conflict, which may account for all the various numbers on water availability in regions where availability has been estimated.

There is a word for those who live their lives relying on guesswork. When you rely on guess-work, estimates and assumptions for your decision-making, we call it carelessness. Estimates and assumptions are not facts. Facts you can rely on; estimates and assumptions may be right or they may be wrong or even almost right.

Being *almost right* is also being wrong.

Groundwater availability estimates, generally, are just that: Estimates. For instance, most all of the studies on aquifers are based on:

    1. *Estimated* average annual recharge (the total of all sources by which an aquifer can be replenished with water);

    2. *Assumed* long-term past and future annual recharge rates and/or amounts;

    3. *Estimated* irrigation use and irrigation return rates;

    4. *Assumed* aquifer complexities;

    5. *Estimates* of aquifer-to-aquifer in-flow and out-flow relationship rates;

    6. *Estimations* or *assumptions* of present and future allocations and use; and

    7. *Judgments* of regional groundwater flow systems,

8. *Predictions* of use and availability in drought conditions, mining scenarios, etc.

The varying results of the numerous aquifer studies have made it difficult for many water-planners and the TWDB to agree on these estimates and assumptions and characteristics of most state aquifers.

This was a factor leading to development of the TWDB's issue of planning guidelines required of the regional planning groups: *"Calculate the largest annual amount of water that can be pumped from a given aquifer without violating the most restrictive physical or regulatory or policy conditions limiting withdrawals, under drought-of-record conditions. Regulatory conditions refer specifically to any limitations on pumping withdrawals imposed by groundwater conservation districts through their rules and permitting programs."*

Even these guidelines require a personal or group opinion as to the interpretation of the term "most restrictive" as it relates to long-term groundwater availability.

And future "regulatory or policy conditions limiting withdrawals" can only be assumed.

The guidelines ignore the complexity of predicting long term groundwater availability, knowledge of all known recharge sources and rates, effect of increasing costs for agricultural and business use, population growth or decline over aquifer areas, and dozens of other factors.

Someone once said that nothing is so dangerous as that which we know, but do not know.

The approach outlined in the guidelines, according to the TWDB, is to conservatively ensure that groundwater resources are not over-allocated. But these guidelines to water planners and those performing aquifer studies, however, are likely to inhibit an independent, accurate study of aquifers in Texas.

When the guy paying your salary tells you what kind of job he wants done, you do it the way he wants it done. Or you better find another job.

It must be recognized and acknowledged that uncertain estimates may exist for many reasons. And foremost of these certainly would include geographical and political reasons.

To demand that water planners "calculate" or predict future ground-water supply when the groundwater demand is unknown is unrealistic. For instance, it is difficult to know under mining scenarios how and when the groundwater in storage will be utilized or the rate and amount of recharge; therefore, it is difficult if not impossible to predict what the available supply will be in the future. Any assumption made is very little different than pure guesswork.

Guesswork may be interesting to consider, but it doesn't further real knowledge.

The concepts of groundwater availability and aquifer sustainability have been debated significantly in recent years.

Forgotten behind the debate is a simple fact: No aquifer has ever gone dry in Texas.

Most of the debate centers around the fact that any estimate may exist for many reasons, including unknown or unknowable complex factors in the aquifer under study: After all, it is underground and, as Courts have stated, unseen.

When it comes to water, the TWDB presents an assumption that offers only two management options:

One option assumes that a static water level in the aquifer will always be maintained, and never exceeded, and the overall water level will not be continually decreased. This is the "safe yield" concept of management.

The second option assumes there will be available long-term water from an aquifer that is equal to the annual recharge volume plus a specified volume of water held in storage within the aquifer. This management scenario is often referred to as "aquifer mining" – in that a water level decline is expected, and the supply will be depleted over time.

Both of the above options have been practiced, in theory, based on the varying hydrogeologic and political factors found in different areas of the state.

For example, aquifer mining has been an accepted policy throughout much of the Ogallala Aquifer in the Texas High Plains because the recharge is relatively low and groundwater demand for irrigation is relatively high. On the other hand, a "safe yield" policy has been adopted for the Edwards Aquifer in Central Texas, partly because of "potential" impact to endangered species that are "assumed" to be dependent on springs discharged from the aquifer.

The fallacy in the TWDB's offer of two management options is the implied, but not stated, necessary control of groundwater by the state.

By rules and regulations, the state controls groundwater districts; and has given oversight of the districts to the TWDB.

The assumption that only a groundwater district or a state unit of some kind with governmental powers is needed for or is capable of management of groundwater is an error in fact and is, in fact, a dangerous precedent.

For the TWDB to use water scarcity or limited available water resources in some areas as a reason to argue for control of all underground water is, at least, questionable.

But isn't control the key factor for the development of a Master Plan on Texas Water Resources?

12 – THE COMING RAID ON NE TEXAS

In a large area of the water-rich region of NE Texas, there is not an underground water district. And the area's property owners still own the water underneath their property.
But the state wants to take their water.

In recent years, the state's water planners have cast their eyes enviously in the direction of Northeast Texas: "*All that water-!*"

Today, decisions about the area's water – and its relationship to Texas' future – are being made in state water circles, without any input by the area's citizens. For this reason, this chapter will concentrate on the state's efforts in this NE Texas area.

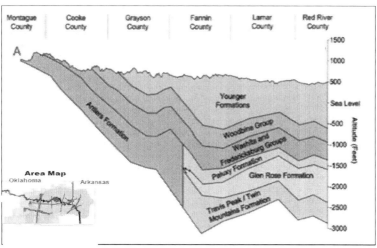

Figure 1

The area is blessed with numerous major and minor aquifers. These include the Trinity, the Woodbine group, which includes the Nacatoch and the Blossom Sands, the Paluxy, and others.

As discussed in the previous chapter, the varying results of numerous studies by various experts have made it difficult for them, the TWDB, and many water-planners to agree on estimates and assumptions.

And this is especially true when dealing with the characteristics of the area aquifers in NE Texas. Many of the factors used to develop "future conditions" of the state's aquifers simply don't apply in this area. For instance, attempts to project the "agriculture use" is laughable, as the area averages over 45-inches of rainfall annually and irrigation is now limited and its use may or may not change.

Rainfall leads to recharge, and other recharge possibilities seem to lead naturally and logically to underground water ways originating under and in mountainous areas of Arkansas and Oklahoma.

With the enormous recharge possibilities, any attempt to "estimate" a rate of recharge is open to question. How can you put water in a pot that is already full?

These questions and others have led to a dizzying and confusing array of opinions and conclusions and arguments by the experts, the TWDB, the Region D Water Planning Group and Groundwater Management Area 8 about the area's groundwater.

The guidelines which the TWDB demands their planning groups use, ignore the complexity of predicting long term groundwater availability, because of unknown recharge sources and rates, and dozens of other factors.

The approach outlined in the guidelines, according to the TWDB, is to conservatively ensure that groundwater resources are not over-allocated. But those guidelines the water planners and those performing the aquifer studies use are likely designed to inhibit an independent, accurate study of aquifers in NE Texas.

Why?

Where else will Texas get the water that Texas needs?

The experts claim Texas will need 27% more water in 2060, and presently has areas claiming a water shortage.

NE Texas is the only area in Texas, where water can be considered a surplus. The water planners know it, and *they want it!*

And for geographical and political reasons are after it.

It must be recognized and acknowledged that uncertain estimates may exist for many reasons. When the guy paying your salary tells you what kind of job he wants done, you do it the way he wants it done. Or you better find another job.

To demand that water planners "calculate" or do prediction of a future ground-water supply when the groundwater demand is unknown is unrealistic, and puts pressure on the one doing the calculation or prediction.

In a real world, any assumption made is very little different than pure guesswork.

Such guesswork may be interesting to consider, but it doesn't further real knowledge.

The guesswork may be right or may be wrong or even almost right.

But to use guesswork to argue for the right to take private property is wrong.

And this wrongness includes more than just estimates and assumptions thrown around in state water circles: The TWDB demands that all water management be done by a governmental unit.

Consideration is not given to the idea that landowners voluntarily joining together to protect, manage, develop and use their underground water would be an effective management tool for their own groundwater.

But chambers of commerce, agricultural associations and cooperatives, from financial products to soybeans to housing, have proven for years that many voluntary organizations can provide effective, successful, responsible management and managers.

Actually, their record beats the record of government.

Shouldn't a particular geographical region determine what is best for it, based on a quantitative understanding of its available resources, whether the resource is water or soybeans or apples or whatever?

Isn't this the basis for a free society?

But in this water-rich area of NE Texas, where property owners have refused to impose a tax or fees on themselves for a water district, the state is attempting to impose one.

Regardless of how area property owners feel, the state has given the Groundwater Management Areas the power to act as a water district. And the area has been placed in GMA 8 - without landowners ever having an opportunity to vote to accept or refuse inclusion in, and control by, a state unit of government.

The right to vote -- the basic building block of American democracy -- on whether or not to place themselves and their property under the control of a groundwater district was denied, and others – outside their area – were given the power to decide for them.

The right for individuals to decide by popular vote should never be taken away by government.

But isn't control the key factor for the development of a Master Plan on Texas Water Resources?

This chapter began with the fact that envious eyes were being turned to a water-rich area of NE Texas. The reason is obvious when just one guess at the area's water resources is examined:

One such estimate, as given by the Region D Planning Group, concludes that 55-million acre feet of "brackish" water exists in aquifer outcrop areas. This is treatable water to potable standards *outside the area's core aquifers*.

Considering that an acre-foot of water contains 325,860 gallons of water, there is more "brackish" water estimated to be in the outcrop areas than there is water contained in some state aquifers. Then, there is the water in the aquifers, themselves. It's a lot of water.

Outside the aquifers, according to the state, there are over 17,908,000,000,000 gallons of just "brackish" water. That's trillions of gallons, which doesn't include the water in the aquifers, nor any recharge.

In April, 2011, T. Boone Pickens sold the water rights under 211,000 acres of West Texas land. The Canadian River Authority, who paid $103 million for the water rights, estimated they purchased around 4 trillion gallons.

Northeast Texas is where the water is . . .

There is no wonder there are those who see the dollars signs in East Texas groundwater. When it comes to politicians and friends in low (and high) places, if there is a buck out there, they think it belongs to them.

Any plan by the state is a plan to take the water in Northeast Texas, as they have no other place to get it.

But, the state does have a lot of votes to satisfy in places that need the water.

13 – GROUNDWATER MANAGEMENT AREAS

(In 2005, the Legislature drastically changed the landscape of groundwater – and groundwater management - in Texas with the passage of House Bill 1763.)

This chapter is alphabet soup – dealing as it does with units of state government. Even as soup its almost indigestible.

Have fun . . .

H.B. 1763 created a new groundwater planning process that requires Groundwater Control Districts (GMAs) to develop joint management plans that establish, in a quantitative manner, the "desired future condition" (DFC) of groundwater resources in each GMA. For example, a DFC could be that water levels do not decline more than 100 feet in 50 years, or spring flow is not allowed to fall below 10 cubic feet per second during a drought of record.

Using the DFC, the Texas Water Development Board (TWDB) must run groundwater availability models that will determine the "managed available groundwater" (MAG) for each district; i.e., how much groundwater is there...

Groundwater "Conservation" Districts (GDCs) must ensure that their management plans contain goals and objectives consistent with achieving the DFC.

Regional water planning groups must use the MAG in their planning efforts. Thus, a district's DFC impacts not only individual property owners in the district, but also the entire planning region.

The joint planning process may allow GCDs to establish *restrictive* DFCs, set a cap on overall production, and *deny permits* once the cap is reached.

For instance, some interpret *Section 36.1132 of the Water Code* as establishing a *prohibition* (or cap) *against issuing permits* in excess of the total MAG established by the DFC.

Once a cap is computed, GCDs may try to create a special, *priority permit* system. GCDs will first allocate MAG to exempt users and historic and existing users, and then divide the remaining MAG among all other users, *many of whom may have been historically conserving groundwater.*

You got all this?
Wasn't it fun?
Government.

14 – THE NE TEXAS WATER PLAN

(The North East Texas Regional Water Plan, recently developed by the Region D Water Planning Group, determined that it is in the best interest of the Region to maintain an acceptable level of aquifer sustainability during the 50-year planning window as well as for future generations beyond the 50-year planning period. Thus, the groundwater availability for the planning period was defined as the amount of groundwater that could be withdrawn from aquifers over the next 50 years that would not cause more than 50 feet of water level decline (or more than a 10% decrease in the saturated thickness – in the outcrop areas - as compared to water levels in 2000.)

After all the words, it is evident that (1) the plan is based on assumptions and estimates, and (2) by allowing all property owners only 10% of the water in outcrop areas, the state takes 90% of all the water in outcrop waters and all the water in the area aquifers. (Outcrop areas are those outside the main aquifer.)

As inaccuracies arise by the use of assumptions and estimates, certain statements in the plan need emphasizing: According to words in the regional plan, *"the criteria used in developing the groundwater availability assessment and to determine groundwater supply for each aquifer in each county, was extremely flexible..."*

Also, *"there were some county-aquifer-basin source groundwater supplies that could not meet the groundwater demands (based on the Region D group's criteria). Therefore, the regional plan states* that *"in those areas, groundwater supply was increased to ensure that all existing groundwater users could continue to use groundwater as a source and potentially expand groundwater use through new strategies."*

Such statements demand questions: (a) If the *estimated* groundwater supplies can't meet the present pumping demand, where is the water now being used coming from? And (b) If, as stated, the *supply was increased,* how was it increased; and from where?

If the supply was increased by the Region D Planning Group's *saying so*, doesn't this simply take care of any water shortages in Texas? Just let the Planning Groups do it by proclamation.

Also, while doing so, they could probably solve the state's budget woes by changing surplus water to wine and opening some state liquor stores.

Such foolishness is no greater than some of the estimations, not facts, that are being used to establish control over the state's property owners.

The planning group acknowledges that *"in some areas, additional water does occur in storage within the aquifers and that a portion of that water (above the estimated supply) could be pumped if there is not a groundwater conservation district in place to prevent such withdrawals."*

Governmental units push for more governmental units. Everything multiplies, except freedom of choice.

Region D, naturally, is pushing for a district. As a state agency, they can do no less; they believe in districts. But the question is, where does this water *"above the estimated supply"* come from; and more importantly, why wasn't it included in the *"estimated supply"*?

It seems the plan contains *estimates of estimates*, which will be the water "Master Plan" for Texas?

According to the Guidance Manual for Brackish Groundwater in Texas, prepared for the TWDB by NRS Consulting Engineers in 2008, there exists *55.8 million acre-feet of brackish groundwater in storage* beneath Region D. (Brackish groundwater contains total dissolved solids content of over 1000 mg/l, and would require treatment to be acceptable for municipal supply. But with 326,000 gallons of water to an acre foot, control over a lot of landowners' water is at stake. See Texas Groundwater Protection Committee chapters.)

Presently in the area, Bowie, Delta, Franklin, Hunt, Lamar, Morris, Red River, Titus, Wood and Rains Counties do not have a groundwater control district. But these counties are under groundwater availability estimates that have been extracted from various reports, and imposed on these counties by Groundwater Management Area 8, the Region D Water Planning Group, and other water planning agencies.

These paid-by-taxpayers' reports include the Desired Future Conditions (DFCs), updated Regional Water Planning pumping estimates (as applied to predictive models), and Groundwater Availability Model (GAM) Runs to determine the amount of Managed Available Groundwater (MAG) for each aquifer.

All of these reports and studies are based on assumptions and estimates that varied according to the source of data and the accurateness of calculation methods, but they all have been accepted by the TWDB. The assumptions and estimates can be changed, however, at the whim of one of the reporting governmental units.

These reports will be used to impose control of the property owner's underground water in area counties where groundwater districts do not exist.

Currently, there is a lot of finger-pointing and ducking for cover:

    (1) The TWDB claims that each GMA "approves their own" Desired Future Conditions;

    (2) The GMAs claim that all they do is (a) determine how much water is currently in aquifers now and how much should be there in 50-years, based on TWDB studies, and (b) that the TWDB is in charge of all "allowable pumping",

exempt and non-exempt wells, permitting, and setting all the rules;

(3) The GMAs and the TWDB all say that it is up to the Regional Water Planning Districts if they want to use these numbers in their planning, but the state says they must;

(4) The Regional Water Planning Groups claim all they do is "use the estimates provided by the GMAs and other TWDB studies" to develop their regional water plan; and

(5) All of the groups in the state use each other to deny responsibility, claiming such things as *"the Legislature authorized us to (whatever it is)"* and *"we have to do what the legislature tells us to do"* and they say, *"If you don't like the way it has been set up, contact your representative …!!!"*

These, of course, are all attitudes which make you love government.

And it is all politics. And it is all about control of the property owner's underground water.

This NE Texas region is under Groundwater Management Area 8 (GMA 8). And the TWDB has told the state's 16 Groundwater Management Areas (GMAs) to develop the "desired future conditions" for all groundwater within their area – which is the same thing they have told the 95-or so groundwater districts in Texas.

This, in essence, gives the same power to the GMAs as that of underground water districts. It must be remembered that underground water districts were established by the vote of the people living within them, but Groundwater Management Areas were created by government fiat.

Basically, this multi-county area is gradually being placed under control of a water district, without a vote: Like the frog in a pot, right now, the heat is no problem: Landowners seem to be happy, waiting to be cooked.

There is controversy over GMAs because the landowner owns the water beneath his property, and as the state is giving GMAs the power to act similarly to a water district this violates the ownership freedom of the property owner. It gives to the GMAs powers of a groundwater district, whether landowners and citizens want it or not. Without allowing a vote of the landowners and citizens in areas without a district it imposes controls on the landowner's use of his or her water.

The State continues to study this issue, though no new legislation was passed by the 79th or 80th legislature, which would protect the property rights of landowners. Bills are pending in this 2011 session offering lip-service to ownership, but which actually give total control of the landowner's groundwater property to districts and/or to the state.

In GMA 8, additional controversy exists because none of these several NE Texas counties have representation on the GMA 8 Board of Directors. It is a board which is composed of several groundwater districts farther west and south.

It is highly unlikely that the interests of these NE Texas counties can be represented by these districts that are 300-and-more miles away. But "local" is the mantra for control.

Why should legislators go out on a limb when they can give the Texas Environmental Quality Commission the power to make up the rules and regulations that will be used to create a groundwater district in this NE Texas area, whether property owners want it or not?

15 – *STEALING APPLES*

Those pushing underground water districts act and think as if they were wearing halos. What they fail to realize is that if the halo were to fall, it would be a noose.

Most of them are long on promises, but short on memory.

Those pushing control are full of *"protect our water"* and *"protect our future"* comments. They keep forgetting: *it's not their water!*

Underground water belongs to the landowner. So, how does it get to be *"our water"*?

And, please, you are gonna' *"protect"* it how? By taking control of the landowner's property? (And they *expect* property owners to *trust* them?)

Again, IF you control it, does it matter who owns it?

Maybe, since water districts can establish their own water fields, take the water, pay landowners nothing, and sell it to their friends, the Land of Oz, or whomever they like, a water district is the preferred method by those who will control the water district?

Those pushing districts depend upon the stupidity of those who see no evil in taking a landowner's property.

As we have mentioned, another thing favored by those pushing districts is establishing water districts by *legislative action* - without a vote of the citizens and property owners within a proposed district.

The point being that the last thing those pushing a water district want is a vote on it.

They claim that if people are allowed to vote on everything, nothing will ever get done; nothing will ever be approved. And people should never have the right to choose for themselves what is okay for them or not okay for them.

Arrogance is when you absolutely know what is best for everyone and absolutely know that everyone should do everything the way you want it done. And you are going to do what you see as the thing that needs to be done in the way that you think it ought to be done.

Water planners are I-specialists. Too often, they are also reprehensible.

The irony, missed or ignored by those asking for legislative action to create a water district, is that the very people they're asking can only achieve office by a popular vote.

Those advocating doing away with our right to vote are contemptible.

The landowner owns the water under his surface acres: There's no *"our water"* to it.

If you have an apple tree on your property, it's your tree, not your neighbors. Nor, is there a neighborhood *"ours"* to it. Certainly there is no right that demands you do a *"sharing evenly"* of the apples falling on your property with a neighbor or other neighbors.

It's your property; they're your apples.

Anyone taking an apple without your permission is stealing apples...

And if your neighbor has an apple tree on his property, there's no *"ours"* to it, either. There is no right demanding a *"sharing evenly"* by him with you or any other neighbor. *It's his property*. If you take his apples without his permission, you are stealing apples...

Underground water belongs to the property owner; it's just more of an asset than apples. So, as of now, no socialists are calling for an Apple Control District. But they are calling for groundwater control…

But there's no difference in principle between stealing apples and stealing water: If a water district has the right to take what is under the owner's property, wouldn't it have a right to take what is on top?

If society allows a right to what is under or on top of someone's property, then it must eventually follow that there is a right, when some greedy little hearts with a lot of money desire it, to the property itself. (Don't ask us to explain stupidity…)

When we start unraveling the Rule of Law, we may lose the thread that leads us out of the maze.

If you're in favor of control of another's property, if you see nothing wrong with taking someone's property, you may be much happier moving to a socialistic nation.

Those seeking control of the property owner's water – yours and your neighbors' – have no right to it.

16 - GIVING LANDOWNER'S HOPE...AND TAKING IT BACK

As previously discussed, water planners and those who seek control of the landowner's underground water now argue that the absolute ownership doctrine does not vest any ownership interest in groundwater until that groundwater is actually produced and reduced to possession.

Under this theory, districts and others, including the legislature, could unreasonably regulate groundwater — in fact, completely halt its production — without any liability for depriving a property owner of that property right.

This is the theory now being heard in the case currently before the Court *(Edwards Aquifer Authority v. Day, 274 S.W.3d. 742 (Tex. App. – San Antonio 2008, pet. pending)*. It will decide whether governmental regulation of groundwater can amount to a *taking* of vested private property requiring compensation.

In the Day case, briefs by the Edwards Aquifer Authority, the Texas Alliance of Groundwater Districts, and the Texas Attorney General have been filed which claim that there is *no ownership* interest until water is actually produced and reduced to possession.

The same argument made in recent years by government, government supporters and water planners.

In real-world terms, however, outside of courts that are becoming more and more politicized, it is an argument to change the terms of the debate from *ownership* to a *vested* right. In real-world terms, it changes a definition – limiting the meaning and intent of property ownership.

As confusing as things are, as confusing as the conflicts within the Texas Water Code, they are made even more confusing by the support of the proposition that the property owner *actually owns the water*, prior to reducing it to possession, by the Texas Legislature in 2003.

That year, through House Bill 803, legislators amended the Texas Property Code to adopt specific procedural and substantive requirements for the condemnation of groundwater rights.

One of the requirements was that the court must consider evidence relating to the market value of the groundwater rights *"as property apart from the land in addition to the local market value of the real property"*, and whether evidence admitted at the hearing shows *"that the real property may be used by the political subdivision to develop or use the rights to groundwater for a public purpose."*

If such findings are made, the court could assess damages to the property owner based on separate considerations of the market value of the real property *and of the groundwater rights*. This would require a variety of specific factors that must be considered in the valuation of groundwater rights.

This treatment by House Bill 803 of groundwater rights, as a component of property to be considered and valued apart from the land itself, is a problem for those seeking control without payment to the owner, and to the court hearing *Edwards vs Day*.

It is entirely *inconsistent* with the idea that the property owner has no compensable ownership right that can be "taken" through groundwater control regulations.

But now, in 2011, some lawmakers are attempting to correct key legislative contradictions. It is a lesson on how a legislative body gives with one hand and takes with the other.

Senate Bill 332, by Senator Troy Frazier (R-Horseshoe Bay), filed on January 12, 2011, was initially thought to strengthen a landowner's ownership of the water beneath his property.

But, by the time the state's Committee on Natural Resources made their additions and deletions (after some suggestions by key Senate members), it was a dangerous piece of legislation for landowners.

After the Committee on Natural Resources' members finished their "corrections", on March 28, 2011, they passed their substitute proposal on for full Senate Consideration by a 12 to 0 vote.

This amended or substitute proposal should be completely opposed by all property owners as it calls for changes in private property ownership and private property rights, as well as more control of private property by government...all in the name of public interest.

This new bill now calls for complete control of all groundwater and, as we have previously warned, sets the stage for control of such districts by the state's 16-groundwater management areas.

While claiming that "the ownership and rights of the owners of the land and their lessees and assigns in groundwater are hereby recognized, and nothing in this code shall be construed as depriving or divesting a landowner the landowner the owners of their lessees and assigns of the ownership interest or rights," it adds, *"except as those rights may be limited or altered by rules promulgated by a district."* (Italics are ours, as are words in parenthesis.)

What rights remain?

If you control it, does it matter who owns it?

The bill is full of doubletalk. It gives with one voice, but takes with another voice. It actually weakens absolute ownership rights in saying that "the legislature recognizes that a landowner has a vested ownership interest in groundwater below the surface *as an interest in the landowner's real property."* (Not absolute ownership, but a vested interest, which ignores the centuries old *ad coelum* doctrines - the common-law rule that a landowner holds everything above and below the land, up to the sky and down to the earth's core, including all of the minerals. [*Black's Law Dictionary 40; 8th edition, 2004].)*

And the bill takes away what it supposedly provides: By including the dangerous provision declaring that "the legislature finds that the conservation, preservation, *use* and development of groundwater resources in this state are *compelling public interests vital to public safety, welfare, and economic progress."* (Basically saying, in plain talk, "As long as we say it is in the public interest, we can take control of anyone's property...")

It also declares that *nothing in the bill shall be construed to*:
- prohibit a district from promulgating *a rule that limits* the right of a landowner to produce groundwater;
- prohibit a district from *limiting or prohibiting* the drilling of a well by a landowner; or
- affect the ability of a district to *regulate* groundwater.

As we ask: If you can control it, does it matter who owns it? Isn't control better than ownership?

If you pay for the property and the costs of ownership, but if I control it, and can take what I want whenever I want it, which one of us has the best part of the deal?

But the taking of ownership rights does end there, as the bill also amends Sections of the Water Code to read as follows:
- A district may *make and enforce rules*...and during the rulemaking process a board *shall* consider *all groundwater uses and needs* (and it is likely their needs will be greater than the need of the individual landowner).
- Consider the *public interest* in conservation, preservation, protection, recharging, and prevention of waste of groundwater...consistent with the objectives of...(the state).

- Consider the goals developed *as part* of the district's comprehensive management plan.
- The presiding officer, or the presiding officer's designee, of each district located in whole or in part in the management area *(a Groundwater Management Area)* shall meet at least annually to conduct joint planning with the *other districts in the management area* and to review the management plans and accomplishments for the management area.

In reviewing the management plans, the districts shall consider:
- The goals of each management plan;
- The effectiveness of the *measures established* by each management plan...and the *effectiveness* of these measures in the management area generally;
- And other matters that *the boards consider relevant. . .;* and
- The degree to which each management plan achieves *the desired future conditions*...

This new, substituted bill demonstrates, as Rural/Urban Resources has consistently warned, that (1) private property ownership rights will be changed or eliminated or subject to controls that allow the state to gain control of all groundwater, and (2) the objective for the state's 16-Groundwater Management Areas is the eventual takeover of all "local" water districts, who will then be used as a "police arm" to enforce the rules and regulations of the state.

Again, we ask, if you can control it, does it matter who owns it?

This bill is a comprise, trying to serve the wishes of the water planners while giving lip-service to landowners. It is dangerous to all property owners, by infringing on all ownership rights, giving to the state more control of an owner's property. It makes a lie of those legislators who claim they are strong supporters of private property rights. It flies in the face of those lawmakers who claim they believe that the powers of eminent domain should be curtailed. It does nothing to control the power of groundwater districts or GMAs but, instead, gives to them much more power.

Generally, it is not the intent of well-meaning men-of-zeal that is dangerous: It is the results that are so much different than the intent.

A tremendous difference exist between the intent of an action and the results. Think of that Noble Experiment, prohibition: The intent of the 18th Amendment was to be in the public's interest; to achieve a *public benefit*. But the results were a proliferation of criminal activity, a strengthening of organized crime, and an increase in the drinking of booze by the American public.

The efforts to re-define ownership rights, to overthrow a century of property rights in Texas, should have everyone, and especially "the men of zeal", asking a few perplexing questions:

- How can the taking of private property rights be in the long-term *public interest*?
- If the state has the right to control groundwater ownership, why wouldn't it have the rights to control the ownership of gravel, sand, or the other minerals *in the ground*, and the rights to control ownership of all the grass, fruit or pecans, and all the timber or crops that grow *on the ground*?
- Isn't destroying rights a dangerous thing to be doing for some perceived *public good, benefit or interest* of a few state water planners?
- When will my property be under attack?
- How much control do we need in our individual lives, and what will be the results of total control? And
- Who will pay for all this control?

If approved, the bill will take effect September 1, 2011.

17 – PLUNDER BY GOVERNMENT

How is taking away a right of a property owner a "good thing"?

Or a thing that honest men should be doing?

To "take" – to plunder – to loot, steal, rob – under law – is a perversion of law, and of the intent of law. But landowners have allowed plunder to become an argument for "we need to control."

Control is to circumvent – to remove the lawful protection of the property owner.

Beginning with Senate Bill 1, the Texas Legislature basically said, We don't care who owns it; groundwater districts are taking control of it!

If the landowner's water is controlled, the use of it taken, does it matter who owns it?

Those we helped elect to office sold out property owners: They gave districts the "regulatory power" to "develop water fields" and "transport water long distances", among other things. And the districts now claim they don't have to pay a dime for the water they pump from or hold in their "water fields".

Legislators have done all this under a "for the public's benefit" excuse.

What is it that lawmakers don't understand about the term, Water Thieves? You change the brand on another man's cows and haul them off to market, it's called "rustling" – and there are laws against it.

But the state can put its brand on the landowner's water and haul it off to market without paying a dime for it?

What is the moral difference?

What the foolish don't understand: The argument is not about water. It's about the individual's right of ownership and the control others want over the rest of us and our property.

"Life, liberty, and property do not exist because men have made laws. On the contrary, it was the fact that life, liberty, and property existed beforehand that caused men to make laws in the first place." (Frederic Bastiat, *The Law* (1850)

"The reason why men enter into society is the preservation of their property." (John Locke, *Second Treatise on Civil Government*)

"When plunder becomes a way of life for a group of men, they create for themselves, in the course of time, a legal system that authorizes it, and a moral code that glorifies it." – Frederic Bastiat, *The Law* (1850)

What Texas property owners should be demanding is that the Supreme Court rule on the use of law to plunder.

Landowners should be arguing that when we start unraveling the Rule of Law, we may lose the thread that *leads us out of the maze.*

18 – THE POLITICALLY BLIND

For those of us who realize that we have way too much government and don't understand why, we need to recognize that too many people are actually kind of dumb or ignorant when it comes to recognizing bureaucracy for what it is when they see it or have to deal with it.

Now, most may actually be semi-intelligent - they're just not able to see reality. A lot of them, however, are ignorant (they just don't know any better).

For instance, they see bureaucracy made up of "public servants", not as an entrenched interest group actively engaged in a systematic effort to look after itself, with no regard for the damage it's doing to you, your finances, your family, and to the country. Looking out for Number One is Number One.

Others are too busy feeding at the public trough to bite the hand that feeds them.

Basically, there are those who are just too dumb to realize what is being done to them - and to others. Some, however, know full well what is happening and are in full support of it: They view themselves as "progressives" or leaders bringing order to ungrateful bunches of people singing "don't fence me in" and who fail to appreciate all the things being done for (to) them –

Bureaucracy is backed by politicians who are backed by the voting masses, and are thereby isolated from any personal responsibility for their own bureaucratic behavior. As they go about their bureaucratic duties to preserve and expand the financial and power rewards of their job, they may smile at you.

Why shouldn't they? The politicians have already given them a firm grip on your money and the power over your freedom of choice.

Those who cannot recognize a bureaucracy for what it is, primarily vote for *moderates* and *progressives* who openly and proudly promise "*a chicken in every pot*" and promote a list of questionable "entitlements" that would gag a hippopotamus. But *moderates* and *progressives* are about all we can find in either major Party, anymore.

Before you get overly excited, don't forget the sorry role Republicans have also played in bringing this nation under bureaucratic control. They have elected their own "*compassionate*" and "*moderate*" towers to political stupidity, offering the same delusions and bringing the same sad results. They just package the programs they're pushing under differently named brands, but they, too, are still pushing "*a small chicken in every pot.*"

Yes, Geraldine, by any name, a "progressive" is a "progressive" is a "re-gressive"!

Once you get past the branding, Democrat or Republican, you'll find the same rush for a nation ruled by bureaucrats forcing socialism - State Capitalism - on us all.

Including those who can't see or won't face reality . . . the dumb and ignorant who are making it possible.

Actually, the problem isn't Democrat or Republican: *It's too many people who fail to recognize socialism run by bureaucracy when it's telling them what to do, how to do it, and when to do it.*

Socialism is the control of production and the means of production by government.

If you have that control, you also control the society where the production is taking place.

This control is placed in the hands of bureaucrats.

In terms of shortening any arguments, try and name three areas of your life where some governmental control does not exist? (You can't do it: Even the air we breathe is taxed. Does the Environmental Protection Agency come to mind?)

Control is control, whether benign or evil.

All control starts out benign. But, the next thing you know, you're up to your rear-end in bureaucrats who don't care a fig about your rear-end or anything else between your nose and your toes. They already have your brain.

And that, my friends and enemies, is real evil.

PART TWO:

TEXAS WATER LAWS

AND THE

TEXAS GROUNDWATER PROTECTION COMMITTEE

Some Background

As in other states, groundwater is vital to the future of Texas.

In 2003, according to the Texas Water Development Board, Texans used about 16 million acre-feet of water, of which 9.3 million acre-feet was groundwater. If correct, groundwater was around 57-percent of all water used.

Of the groundwater used, about 40-percent was for irrigation, with the remainder used for municipal, rural and domestic consumption, livestock, electric utility, and industry.

Approximately 36 percent of the municipal water used was obtained from a groundwater source.

Major and minor aquifers underlie approximately 76-percent of the state's 266,807 square mile surface area.

Major aquifers are defined as producing large quantities of water in a comparatively large area of the state, whereas minor aquifers produce significant quantities of water within small geographic areas or small quantities in large geographic areas.

Minor aquifers are important because they may constitute the only significant source of water in some regions.

The Texas Water Development Board (TWDB) has delineated nine major aquifers and 21 minor aquifers. Current maps of the aquifers are available on the TWDB's web site at:
<www.twdb.state.tx.us/mapping/index.asp>.

In some areas of the state, "undifferentiated" local aquifers may represent the only source of groundwater where major or minor aquifers are absent. These local aquifers vary in extent from being very small to encompassing several hundred square miles.

All these aquifers, or portions of them, are under the surface of land owned by private property owners – who own the water beneath their land.

These aquifers contain an immense amount of water, which equals out to a lot of potential money and/or power for the owners.

Government does not become an aggressor until it discovers something it wants that belongs to another. Then, it marshals forces, organizes a plan of attack, and takes action to acquire what it wants. As King Arthur observed, "Might has little to do with Right."

Today, property owners are facing an intimidating number of cooperating governmental agencies and their dependent pawns, forces engaged in battle plans designed to take control of the landowners' water, thereby usurping private property ownership rights.

It is an expensive war. For instance, the 81st Session of the Texas Legislature added around $170-million to the budget of the *Texas Water Development Board.*

Property owners are paying them to take control.

The state has more attorneys on staff than most individuals can afford to hire; there are so many lobbyists, publicity pushers, consultants, advertising gurus and campaign managers pushing water control they're bumping into each other; all the politicians are figuring out ways to get what they want; and the state, by far, has *more guns, and bigger ones, too.*

Everyone is scheming ways to get what they want. That is, everyone but the property owners.

It is a stealthy war, and for too long a period most property owners were not aware they were under attack. And too many still don't know it or believe it.

But arrayed against them is an imposing list of formidable adversaries, which include:

The Texas Commission on Environmental Quality (TCEQ);
Railroad Commission of Texas (RCT);
Texas Department of Agriculture (TDA);
Department of State Health Services (DSHS);
Texas State Soil and Water Conservation Board (TSSWC);
Texas Department of Licensing and Regulation (TDLR) (water well construction);
Texas Water Development Board (TWDB); and the
Texas Association of Groundwater Districts (TAGD).

Within and associated with these state agencies and boards are numerous other sub-agencies, offices, organizations, committees, etc., which have been assigned the responsibility and the authority for doing research and fostering educational programs and publicity, dealing with groundwater issues statewide.

It is almost impossible to determine the total number of state and federal agencies involved with underground water, but all approach control with an aggressive zeal, using "protection" as their excuse to do so.

A few short years ago, it was all about "conserving" water. But, people began asking, "Conserving it for whom?"

So, the role of each state agency expanded: Now, in addition to "conservation", it is also about the *need* for "protection" and "prevention". Not just in regard to water, but for the "health" of all the state's citizens.

It all sounds good and noble and the thing to be doing.

But is the state really "protecting" citizens by taking their ownership rights? Whatever the intentions, will the results be good and noble, and the thing that honest men and women should be doing?

Unfortunately, government rules by might.

Control has come (and continues to expand) a little bit at a time, incrementally.

Policies, guidelines and *recommendations* have been formulated piecemeal, slowly approved and accepted as *"the state's policy"* or as an *"agency desired way"* of doing a particular thing. Enough SB 1's, SB 2's and SB 3's have been enacted to establish *the process* and *the intent* and *the meaning* of the law.

If it were a good thing to be doing, why do it so piecemeal?

The High and the Mighty

Government sells bad things with good words. And government uses a lot of words.

Government prints a lot of stuff, and you have to read between the lines to get a faint glimpse of what all the words really mean.

Here are two of the most powerful agencies leading the takeover of property ownership rights.

There are a lot of *protects, develops, conserves, preserves, authority, provides, environment, regulatory, permitting, prevention, enforcement, implementation, responsibilities, quality, problems, management, remediation,* and *programs* in the objective and goals of agencies dealing with groundwater. Unfortunately, all of these words, and more, can be defined as being related to *control*.

A few of the hurdles now before property owners include the great-sounding words of the:

<u>Texas Commission on Environmental Quality.</u>

The TCEQ *claims responsibility* for the majority of the state's *environmental* and water *quality regulatory* programs.

It conducts a number of *programs* that address groundwater control, while publicly focusing on "*prevention* of contamination" and *remediation* of "existing *problems* of water quality."

The TCEQ *implements* these *programs* through education, voluntary action assistance, *permitting*, and, not least of all, *enforcement*.

As the lead state agency for water *quality* and *environmental protection*, the TCEQ administers both *state and federally mandated programs*. Federal *programs* include:
- The Resource *Conservation* and Recovery Act for the management of municipal and industrial wastes;
- The Comprehensive *Environmental* Response, Compensation, and Liability Act or Superfund *environmental* cleanup program;
- The Clean Water Act for *managing* pollutant releases to state waters;
- The Safe Drinking Water Act for the *protection* of public drinking water supplies; and
- The *development* of a PMP for the *protection* of groundwater under the Federal Insecticide, Fungicide, and Rodenticide Act (FIFRA).

Plus, the TCEQ claims *responsibility* and *authority* under state law provided in the Texas Water Code and the Texas Health and Safety Code for a number of *programs* addressing water resource *management*, waste *management*, and *environmental protection*.

(You name it, these boys and girls can do it!)

The TCEQ is headed by a three-member commission and is organized into several major functional *programs*:

The Office of *Permitting* and Registration, and the Office of Water are *responsible* for *permitting* facility operations that include provisions to *prevent* groundwater impacts, and for *providing* support to the TGPC.

The Office of Compliance and Enforcement is *responsible* for assuring that *regulated* entities comply with *permits* and agency rules including provisions related to groundwater *quality protection* through a network of agency regional offices, facility inspections, *enforcement* proceedings, professional licensing; for *remediation* and *corrective* action to address groundwater contamination; and *implementation* of the Edwards Aquifer *Protection* Program.

The Office of Water is also responsible for *developing* and *implementing* plans for *achieving* clean water.

Programs throughout the TCEQ *provide* outreach and technical assistance to specific stakeholders and *regulated* communities.

The TCEQ also has outreach *programs* designed to help small businesses and local governments.

As stated, you name it and these boys and girls can do it.

Whether or not you want it, and even if it is not needed nor wanted.

Then. . .there is the *Texas Water Development Board (TWDB)*.

The TWDB was created in 1957, and is the state agency that is *responsible* for *statewide water planning*; collection and maintenance of water resource information; and *administration* of financial assistance *programs* for water supply, water *quality*, flood control, and agricultural water *conservation* projects.

The TWDB is *responsible* for the *development* of the State Water Plan to *provide* for the orderly *development, management*, and *conservation* of the state's water resources. The TWDB provides support to Regional Water *Planning* Groups for the *development* of regional water plans that are used to prepare the State Water Plan.

The TWDB, in support of its water planning and data collection *responsibilities*, conducts an active groundwater resource assessment program. The TWDB conducts studies to assess the state's aquifers, including occurrence, availability, *quality*, and quantity of groundwater present. It also identifies major groundwater-using entities and current and projected demands on groundwater resources.

The TWDB conducts statewide groundwater level measuring and groundwater *quality* sampling *programs* as a part of its assessment effort. The groundwater quality-sampling *program* requires the TWDB to: (1) monitor changes, if any, in the ambient *quality* of groundwater over time, and

(2) establish the baseline *quality* of groundwater occurring naturally in the state's aquifers.

As a significant part of the water planning process, the TWDB supports the *development* of Groundwater Availability Maps (GAMS), which are state-of-the-art, publicly available numerical groundwater flow models. GAMs are tools to help in the process of *determining* groundwater availability in Texas in order to ensure adequacy of supplies or recognition of inadequacy of supplies throughout the State Water Plan's 50-year planning horizon.

The TWDB has 20 models covering all nine of the state's major aquifers and several of the minor aquifers, and is now working on *developing* 12 additional models for the remaining minor aquifers as well as updating existing models to include new aquifer and water use information.

Furthermore, the TWDB is charged with providing *Managed* Available Groundwater (MAGs) amounts along with other technical assistance to Groundwater Management Areas in support of the desired future conditions process *mandated by House Bill 1763*, signed into law in 2005.

The bill required the sixteen GMAs around the state to set desired future conditions of the aquifers in their jurisdiction by September 1 of 2010. *(Or the TWDB would do it for them…)*

The groundwater technical assistance and the groundwater modeling areas have been providing assistance in a variety of formats to these planning entities.

Once the GMAs *determine* their desired future conditions, staff from the groundwater division of the TWDB will provide *managed* available groundwater numbers to the GMAs as well as the regional water planning groups.

As you can tell, it just doesn't get any better than the TWDB. As described, mostly in their words, how can you not appreciate their work?

While it is believed the Texas Commission on Environmental Quality (TCEQ) is the "hatchet man" in state water circles, it is the TWDB that holds the real power.

A law passed quietly by the legislature that nibbles ever so gently at groundwater ownership rights is never one pushed by the TWDB, as far as public knowledge is concerned. It just happens or it is one sponsored by another agency.

Like Lola, *whatever the TWDB wants, the TWDB gets*.

Beyond any reasonable doubt, the TWDB reports well.

They can make a property owner *want* to give up their rights.

Other Major Players

Some of the other players that wield considerable influence and power in the water battle lines include:

>*The Railroad Commission of Texas (RTC).* The RTC's regulatory authority includes oil and gas exploration and production, surface mining and mine reclamation, and pipelines, along with regulatory responsibilities.

Permits to drill oil, gas, and related wells are issued only after the applicant has submitted a letter from the TCEQ that provides information on the depth of usable quality groundwater.

Knowledge of the presence of shallow groundwater and the recharge areas of aquifers is vital to the regulation of surface storage and disposal of oil and gas wastes. (Underground injection including hydrocarbon storage, and brine mining, are primarily groundwater protection regulations federally delegated under the Safe Drinking Water Act.)

A groundwater impact assessment is performed as part of the surface mining permitting process. *Permits* contain plans to protect the groundwater resources in the area of the *permit*. Groundwater may be removed during the mining activities; however, if those activities adversely impact a currently used groundwater resource, then the impacts must be mitigated.

The Texas Department of Agriculture (TDA) is the lead authority for pesticide regulation in Texas. TDA evaluates the potential impacts of pesticides on human health and the environment, including groundwater and surface water *quality*.

TDA staff participates on all interagency subcommittees and task forces charged with various aspects of groundwater protection. The staff participates in statewide, regional, and local regulatory and voluntary programs and committees focusing on water quality, water supply, conservation, and other issues which influence or impact groundwater use in the state.

> *The Department of State Health Services* (DSHS), formerly the Texas Department of Health, has limited involvement in groundwater protection, although it does provide services that are related to groundwater safety and public health concerns. With regard to groundwater issues, the Community Hygiene Group in the Division of Regulatory Services acts primarily in a non-regulatory manner and serves in an advisory or public service role.

Although they have no direct programs that relate to groundwater protection, the DSHS does have programs that indirectly provide protection to the state's water resources. Under the Regulatory Licensing Unit, the Chemical Reporting Group administers and enforces Tier II reporting of hazardous substances. The Policy Standards and Quality Assurance Unit oversees programs for youth camps, childcare centers, and investigates public health nuisance complaints.

The DSHS Laboratory Services Section performs and works under contract with other state agencies such as the TCEQ.

> *The Texas State Soil and Water Conservation Board* (TSSWCB) was created by the Texas Legislature in 1939, to organize the state into soil and water conservation districts (SWCDs) and to serve as a centralized agency for communicating with other state and federal entities as well as the Texas Legislature. It offers technical assistance to the state's 217 SWCDs and maintains regional offices in strategic

locations in the state to help carry out the agency's water quality responsibilities.

As its members are scattered over a large area of Texas, living in many communities, the TSSWCB wields a tremendous amount of influence in selling the state's water plans.

The TSSWCB also works with other state and federal agencies on issues as they relate to Water Quality Standards and Criteria, Total Maximum Daily Loads, and Coastal Zone Protection.

The TSSWCB has authority to establish water quality management plans in areas that have developed, or have *the potential* to develop, agricultural or silvicultural water quality problems.

> *The Texas Alliance of Groundwater Districts (TAGD)*, formerly the Texas Groundwater Conservation Districts Association, was formed on May 12, 1988. Its core membership is restricted to groundwater *conservation* districts in Texas, with the power and duties to manage groundwater (as defined in Chapter 36, Texas Water Code). Other organizations with an interest in groundwater management may become an Associate Member.

TAGD is a non-governmental organization, and has no regulatory or enforcement authority. However, Groundwater Control Districts (GCDs) that participate in TAGD have authority *over* groundwater use and contamination.

TAGD is organized exclusively for charitable, educational, or scientific purposes within the meaning of Section 501 (c) (3) of the Internal Revenue Code. As such it can accept tax deductible donations and use these donations to educate the public to the growing need for water conservation and groundwater protection.

The purpose of TAGD is to educate the public, further groundwater conservation and protection activities, and to provide a communications vehicle for the exchange of information between individual districts and the general public.

TAGD maintains contact with members of the private sector and various local, state, and federal officials and their agencies to obtain, and provide, timely information on activities and issues relevant to groundwater conservation districts.

Districts are created by the legislature or by the TCEQ with the purpose of controlling groundwater. Such districts can be created by one of three procedures:

(1) Special law districts can be established by the legislature (introduction and passage of a bill to do so);

(2) Districts can be created in priority groundwater management areas through procedures initiated by the TCEQ (Sections 35.012(b) and 36.0151 TWC). No vote required. And

(3) Districts can be created through a property-owner petition filed with the TCEQ (Section 36.013, Texas Water Code). Generally, after a majority vote approval by those living in the boundaries of the proposed district.)

Districts are local or regional in their jurisdiction.

> *The Texas AgriLife Research* is the *official agricultural research agency* in Texas. Goals of the AgriLife groundwater research program are to *protect, preserve,* and efficiently use water resources and to *develop* sustainable agricultural production systems.

Groundwater programs of AgriLife Research stress the development of management strategies, technologies, and educational programs to support sustainable agriculture and related natural resources management.

Texas AgriLife Research trains future professionals through undergraduate and graduate education and research programs at Texas A&M University and other System institutions.

AgriLife Research efforts are complemented by the programs of the Texas AgriLife Extension Service, also a component of the Texas A&M University System.

AgriLife Extension conducts educational programs on management strategies to *protect* groundwater resources.

> The Bureau of Economic Geology (BEG), established in 1909, is a research entity of The University of Texas at Austin and functions as the *State Geological Survey*. BEG is one of three member institutions within the School of Geosciences. One of its goals is to conduct research related to water issues in Texas. The BEG conducts such basic and applied research projects in support of other state and Federal agencies. The BEG is not a *regulatory* agency and has no groundwater *protection regulatory* programs but supports the agencies that fulfill these functions.

The BEG serves as a valuable resource for geologic maps and reports that provide the framework for many environmental studies. One of the strengths of environmental studies conducted by the BEG is the integration of geology and hydrology. Groundwater resources are the focus of several studies.

> *The Texas Department of Licensing and Regulation* (TDLR) maintains the Water Well Drillers Advisory Council; investigates all alleged violations of Chapters 1901 and 1902 of the Texas Occupations Code and 16 TAC Chapter 76 (Water Well Drillers and Pump Installers Rules); investigates consumer complaints filed against regulated well drillers/pump installers; and randomly inspects wells to insure compliance with well construction standards.

The program also works with federal and numerous state and local entities in the area of groundwater protection.

Behind the Lines

In any war, ammunition is needed. The more ammunition on hand, the better is the likelihood of winning battles, and the war itself.

Words are the ammunition in the Texas Water War.

If you can convince the populace that your taking the property of others is for their benefit, most will support you.

Remembering the keywords that relate to the word control, here is the – whatever it is – that provides the excuses for the war:

In 1989, the Texas Legislature created the The Groundwater Protection Committee to *"bridge gaps and improve coordination among existing state water and waste regulatory programs."* The Texas Water Code [§26.401 through 26.407] established the TGPC, and outlined its powers, duties, and responsibilities.

When creating the TGPC, the legislature charged it with the responsibility to and an objective of developing and updating a "comprehensive" groundwater protection strategy that would include guidelines for the prevention of contamination, the conservation of ground-water, and the coordination of the groundwater protection activities of the agencies and entities represented on the TGPC.

(If you are remembering to "read between the lines", the emphasis is on *"the conservation of groundwater, and the coordination of...protection..."*)

The TGPC currently consists of nine state agencies and the Texas Alliance of Groundwater Districts.

Over the last biennium, the TGPC continued to use existing policy and the programmatic direction given by the legislature as the basis for implementation of its activities. The *Strategy provided by the Legislature* also provided recommendations and possible actions that were to be taken over the period of time between 2003 and 2013 to enhance protection of groundwater.

But entering 2000, the TGPC was in full swing.

With the continuing state focus on the need for assuring a high quality supply of groundwater, and recognizing the programmatic changes that had occurred since the state's first groundwater protection strategy was developed in 1988, the TGPC decided in January 2001 to update the state's groundwater strategy.

The TGPC issued the revised *Strategy* in February 2003, which contained the implementation of the following objectives:

- Agricultural Chemical Activities;
- Groundwater Data Management Activities;
- Nonpoint Source Pollution Activities;
- Public Outreach and Education Activities;
- Groundwater Research Activities;
- Intergovernmental Cooperation Activities; and
- TGPC Administrative Activities.

In creating the TGPC, the legislature established a policy of non-degradation of the state's groundwater resources as the goal for all state programs. *(Which in itself, is a Hell'ov an excuse to take over groundwater.)*

According to the TGPC, the state's groundwater *protection* policy recognizes:

- the variability of the state's aquifers in their potential for beneficial use and susceptibility to contamination;

- the value of *protecting* and maintaining present and potentially usable groundwater supplies;
- the need for keeping present and potential groundwater supplies reasonably free of contaminants for the *protection* of the environment and public health and welfare; and
- the importance of existing and potential uses of groundwater supplies to the economic health of the state.

The TGPC recommends the use of the best professional judgment of responsible state agencies in attaining the goals and policies of the *strategy. (And why wouldn't they?)*

The TGPC claim they implement policies by identifying opportunities to improve existing groundwater quality programs and promoting coordination among agencies.

(Which would seem to mean that the TGPC looks for excuses to recommend areas where new control programs can be started or existing programs enhanced "to provide additional protection.")

The major responsibilities claimed by the TGPC are to:
- improve coordination among member agencies and organizations engaged in groundwater protection activities;
- develop, implement, and update a comprehensive groundwater protection strategy for the state;
- study and recommend to the legislature groundwater protection programs for each area in which groundwater is not protected by current regulation;
- file with the Governor, Lieutenant Governor, and Speaker of the House of Representatives a biennial report of the TGPC's activities and any recommendations for legislation for groundwater protection;
- publish an annual groundwater monitoring and contamination report describing the current monitoring programs of each member agency and the status of groundwater contamination cases documented or under enforcement during the calendar year; and
- advise the TCEQ on the development of plans for the protection and enhancement of groundwater quality pursuant to federal statute, regulation, or policy, including management

plans for the prevention of water pollution by agricultural chemicals and agents.

In short, it would appear that the job of the TGPC is to manufacture the ammunition used to take control over the ownership rights of the state's property owners.

Periodically, state laws have been enacted that require the TGPC to undertake rulemaking – farming out the responsibility for law-making to the committee.

Today, the TGPC is one of the most influential bodies in Texas, as seen in its preparedness for the convening of the 82nd Legislature.

In January, 2011, the TGPC submitted a prepared list of activities and recommendations concerning groundwater to members of the 82nd Session. (Copies are available through the Texas State Library, as well as other sources.)

State law (Texas Water Code [TWC]), as written and passed by legislators, supporting state water planners, requires that the state water planners make reports providing reasons (excuses) worthy of the lawmakers' support.

These reports, in return, give the lawmakers something to point to as their reasons for support of the water planners. It is a neat, closed, circle. Each member of the "circle" can point to the other members of the circle to justify whatever it is that each wants to do . . .

Politics at its finest.

The report submitted by the TGPC to the 82nd Texas Legislature was done *as required* by state law [Texas Water Code (TWC) Section 26.405]. It provided recommendations to improve groundwater protection for legislative consideration, and described the TGPC's activities for the preceding biennium.

The following content is taken, in part, from the TGPC report, as presented to the legislature, and without editorial comment:

ACTIVITIES AND RECOMMENDATIONS OF THE TEXAS GROUNDWATER PROTECTION COMMITTEE: REPORT TO THE 82ND LEGISLATURE:

Twelve groundwater protection recommendations are presented in this report requesting legislative consideration in three topical areas:
Strengthen groundwater conservation and water quality protection efforts,
Advance groundwater management and protection through enhanced data collection and availability, and
Support groundwater research.
The recommendations are as follows:

Protect Groundwater Quality through Education Programs:

Issue. To best protect groundwater quality, a variety of education programs are needed to transmit information to the public. These programs will provide resources for water resources managers and agency personnel to demonstrate use of innovative technologies and management strategies.

Recommendation. Support ongoing groundwater education, demonstration, and outreach efforts administered by the Texas AgriLife Extension Service and TWRI. The efforts would be coordinated with the TGPC and other entities.

Background. Several groundwater education programs are already in place, led by the Texas AgriLife Extension Service and several other agencies and entities. Some of the broad topics addressed by these education programs include the following:
• Protection of drinking water wells and areas where wellheads are located;
• Ways to ensure that wellheads are not contaminated and abandoned wells are properly plugged;
• Proper selection, use, and management of on-site wastewater treatment systems;
• Pesticide laws, regulations, and actions agricultural producers can take to reduce pesticide use and limit the risk of degrading groundwater quality; and
• Demonstrations that show how to plug abandoned wells and how individuals can take groundwater quality samples.

These educational programs need continued financial support to achieve long-term results. In addition, new efforts needed to address issues include:
• Preparing individuals and groundwater suppliers to deal with threats to water quality;
• Explaining the economic benefits of protecting groundwater quality;
• Encouraging stakeholders to participate in discussions about current groundwater quality and quantity issues and help identify future water needs;
• Identifying new and emerging technologies that have a significant potential to treat and remove groundwater contaminants; and
• Providing urban, suburban, and rural-fringe education:

Ensure that Brackish/Saline Aquifers Having Potential for Use as Drinking Water Are Protected from Contamination:
Issue: TWC, §26.401 stipulates that, in order to safeguard usable and potentially usable groundwater, it is the policy of this state that discharges of pollutants, disposal of wastes, or other activities subject to regulation by state agencies be conducted in a manner that will maintain present uses and not impair potential uses of groundwater or pose a public health hazard. Brackish groundwater and certain saline groundwater is gaining importance as a source of drinking water now and in the future. Regulatory programs may not have anticipated the potential use of brackish and certain saline water for human consumption, and therefore may not have provided adequate protection of the resource from pollutant discharge or other contamination.

Recommendation: Encourage state regulatory agencies to examine existing and proposed policies and rules to ensure that brackish and saline groundwater sources, identified as having potential use as drinking water, are adequately protected from contamination.

Background: Brackish groundwater is defined as groundwater containing between 1,000 and 10,000 milligrams per liter (mg/L) total dissolved solids (TDS), while saline water is defined as having a TDS content of greater than 10,000 mg/L. Brackish and saline groundwater can be found throughout all of the regions of the state.

As water demands increase and freshwater supplies decrease, more cost-effective desalination technologies are allowing for widespread use of the resource for drinking water supplies.
Currently, there are approximately 100 public water systems in Texas using desalination to treat nearly 80 million

gallons of water per day and, according to the 2007 State Water Plan, 3.5 percent of the new water supplies to be developed by 2060 will be provided by desalination. Brackish groundwater sources for desalination were identified in a TWDB report as to location and amount available for desalination. The total estimated volume of brackish groundwater "in place" in Texas aquifers is over 2.8 billion acre-feet.

Current regulatory practices, including risk reduction programs, generally afford slightly less protection to brackish or saline groundwater than they do for fresh groundwater supplies. Rules and policies need to provide protection necessary to maintain suitability of brackish and saline groundwater that is identified as having potential for use as drinking water, for cost effective desalination treatment.

Advance Groundwater Management and Protection through Enhanced Data Collection and Availability:

To ensure the best management of the state's groundwater supply, local and regional planning groups must develop approaches and management methodologies based on high-quality groundwater data; real-time groundwater data; information developed from the completion of groundwater availability models (GAMs) for all of the state's minor aquifers; and sound, defensible determinations of desired future conditions, and calculations of managed available groundwater. Existing data need to be captured into a user friendly database.

Support the Statewide Real-Time Groundwater-Level Monitoring System:

Issue. Texans need real-time water-level information to manage their groundwater resources. Such information helps

Regional Water Planning Groups (RWPGs) and water suppliers develop drought management plans and individual well owners understand current conditions within an aquifer. Groundwater Conservation Districts (GCDs) and interested citizens depend on real-time water-level information from automatic recorder wells to monitor day-to-day changes in water levels. More GCDs are using data from these recorder wells to determine different drought management stages, as the Edwards Aquifer Authority has, using data from the San Antonio J-17 index well, for more than a decade in managing pumpage from the Edwards (Balcones Fault Zone) Aquifer in Bexar County.

Real-time water-level information is obtained by equipping a well with automated water-level measuring or recording equipment and a transmitter to send information to a central location for posting on the internet. The TWDB current real-time monitoring network has 150 monitoring stations in 70 counties. The TWDB's interactive map also links to web sites of other agencies with recorders that also publish real-time water-level measurements. However, the current network consisting of recorders from all agencies is inadequate for assessing all of the state's groundwater resources.

Although the TWDB strives to add recorders with telemetry equipment to counties in need as the agency's yearly budget allows, currently nearly 50 predominantly single-county groundwater conservation districts and nearly 100 counties with no districts do not have the necessary resources to monitor groundwater levels and host online, real-time water-level information.

Provide continued support to the TWDB's baseline budget which currently allows for a dozen additional sites every two years, or an even greater number of sites when the agency partners with cooperating entities that are able to provide partial funding. An expanded network is needed to achieve parity in the geographic distribution of all real-time monitoring sites and provide all counties with at least one real-time recorder to complete the network. Groundwater districts that are able to partner with the TWDB will benefit from the additional information such recorders can provide in managing their groundwater.

Background. The TWDB has operated recorders throughout the state for several decades. In the past decade, automatic recorder data have become available on a daily basis, on demand, through the installation of dataloggers and telemetry that allow posting of data on the TWDB's web site. The legislature's passage of Senate Bill 1 funded initial TWDB efforts to launch a real-time recorder program and publish water levels online. The agency continues to operate and enhance this program through purchase of additional satellite telemetry and recorder equipment. Other groundwater conservation districts and universities also are publishing their real-time data on the TWDB web site, particularly those districts that have been able to purchase equipment that the TWDB then helps install and maintain. The continued development and maintenance of this program will allow for the purchase, installation, and maintenance of recorders in all areas of Texas and the dissemination of this information to the public in real time.

Continue Support of "Desired Future Conditions" Process:

Issue. House Bill 1763, enacted by the 79th Legislature in 2005, requires GCDs to determine the "desired future conditions" of their groundwater resources by September 1, 2010, and the TWDB to provide estimates of managed available groundwater to the districts and the RWPGs. People with defined interests in groundwater can petition the TWDB if they believe that the "desired future conditions" determined by groundwater conservation districts are not reasonable.

Recommendation. Continue to support TWDB's implementation of House Bill 1763.

Background. With House Bill 1763, the 79th Legislature greatly expanded the role of groundwater management areas in managing the groundwater resources of Texas. Groundwater conservation districts in each of the sixteen groundwater management areas now are required to meet to decide the "desired future conditions" of their groundwater resources. The "desired future conditions" then are used to calculate the "managed available groundwater," which is the amount of groundwater available for permitting and the amount of groundwater available to meet future demands in regional water planning.

The process of deciding "desired future conditions," calculating "managed available groundwater," and responding to petitions against desired future conditions requires considerable technical and legal support, especially if the state desires defensible numbers. When House Bill 1763 was being considered, the TWDB submitted a fiscal note, approved by the Legislative Budget Board, which included additional employees to implement the bill. But House Bill 1763 was approved toward the end of the session and there was insufficient time to consider appropriations for the fiscal note.

With legislative appropriations in 2009, however, the TWDB was authorized and funded to hire employees to implement the program.

Characterize Groundwater Surface Water Interactions:

Issue. The TCEQ 2008 Water Quality Inventory identified numerous stream segments that are impacted by biological contaminants.

Surface water, particularly located in urban streams, is often contaminated by both chemicals and pathogens, disease-causing organisms such as bacteria, which can cause illnesses even at low concentrations. Potential pathogen sources include septic tanks, public wastewater treatment plant effluent, land application of sewage sludge and leaking sewage collection systems. Groundwater is also vulnerable to pathogens from surface sources where rapid infiltration or limited filtration capacity occurs. Groundwater in karst systems is particularly vulnerable to pathogen contamination from surface sources, as a result of rapid infiltration via sinking streams, open fractures or sinkholes. However, the extent of pathogens in groundwater as a result of nonpoint sources, and specifically pathogen persistence, concentrations and modes of transport potentially from surface water to groundwater, has largely not been investigated.

Increasing demands on both surface water and groundwater resources will increase the implementation of treated wastewater reuse, recycling, dual water systems and aquifer storage and recovery systems; all of which provide additional opportunities for groundwater and surface water to become adversely impacted by contaminants because of the interrelationships between these interdependent sources of water.

Recommendation. Provide support for investigations to address pathogen residence time, survivability, rate of transport, and methods of transport of viruses, bacteria, protozoa, and other contaminants from surface water to groundwater and from groundwater to surface water.

Background. Groundwater and surface water are intimately related within the hydrologic cycle. Streams receive inflows from groundwater discharge through stream banks and streambeds and groundwater can thereby influence the quality of surface waters. The reverse is also true; discharge from streams during periods of high stream stages result in infiltration via bank storage, and recharge to riparian groundwater aquifers, and potentially can impact groundwater quality.

Drinking water supplies derived from wells have, in general, been assumed to be relatively safe from chemical or pathogen contamination because of the filtration capacity of surface soils and the unsaturated zone above the water table and the ability of aquifers, as porous media, to filter out biological and some chemical contaminants. Where concerns of adverse impacts of surface sources of pathogens on groundwater have arisen, the solution has often been to ensure that surface casings for wells are adequately sealed and of sufficient length to isolate the production zone of the well from direct infiltration of surface water. Regulations governing the required distance of septic tanks and drain fields from water supply wells are an example of this approach.

Groundwater is vulnerable to chemical and biological contaminants from surface and subsurface sources where rapid infiltration or limited soil filtration capacity occurs. However, the extent of pathogens in groundwater as a result of nonpoint sources and specifically pathogen persistence, concentrations, and modes of transport, has not been investigated in great detail.

Where the direct relationship of groundwater impacts resulting from contaminated surface sources has been investigated, it has most commonly been focused on chemical contaminants and not on pathogen movement from sources to groundwater. Much work remains to be done to define the controlling factors that affect pathogen impacts on groundwater quality.

Overview of the *Groundwater Protection Strategy:*

In developing the *Strategy*, the TGPC recognized that the state has numerous successful groundwater programs spread among local and state governmental agencies and research institutions. Therefore, a key part of the *Strategy* documented how the current regulatory, outreach, and research programs work to protect groundwater resources. A second component of the *Strategy* was the identification of protection gaps in program implementation or coordination. TGPC believes that the *Strategy*, grounded firmly within the existing policy and programmatic directions given by the legislature, resulted in a document that sets realistic objectives for success and provides a road map for action over the next five to fifteen years.

The *Strategy*:
- details the state's groundwater protection goal as established by the legislature;
- explains the state's efforts to characterize the occurrence, quality, and quantity of groundwater resources and discusses various assessment approaches used in program implementation;
- describes the roles and responsibilities of the various state agencies involved in groundwater protection and discusses the TGPC as a coordinating mechanism;

- provides examples of how the various state agencies implement groundwater protection programs through regulatory and nonregulatory models;
- explains how local, state, and federal agencies coordinate management of groundwater data for the enhancement of groundwater protection;
- discusses the role that research plays in understanding groundwater's importance and the importance of coordinating research efforts;
- provides an overview of groundwater public education efforts in the state;
- discusses public participation in establishing and implementing groundwater policy;
- lays out a planning process for updating the *Strategy*;
- proposes for inclusion in the next *Strategy* an identification and ranking of significant threats to the state's groundwater resources, consideration of the vulnerability of groundwater resources to such threats, and a prioritization of actions to address those threats; and
- provides recommendations and possible actions to protect groundwater.
- The TGPC needs to strengthen the lines of communication and information sharing with the State's RWPG. The lack of communication between these two programs is a gap in the TGPC's ability to coordinate the state's groundwater protection strategy with the state's water supply planning efforts led by the TWDB.

Improve Groundwater Data:
- The existing groundwater quality monitoring programs need more resources to sample additional sites that will provide a better picture of groundwater conditions statewide.
- The parameters that are analyzed need to be expanded to include organic and synthetic chemicals. While site-specific assessment of hazardous wastes in groundwater is covered by

a number of state and federal programs, other substances in groundwater, such as nitrate and arsenic that may be deemed naturally occurring, need better assessment.

- The TGPC should develop recommendations on the design of a groundwater monitoring system that will meet the needs of all member agencies and organizations. Any new monitoring of domestic water wells would be on a voluntary basis.
- Data management standards should be periodically reviewed and amended to facilitate information exchange. The TGPC must review and revise its groundwater data management standards and guidelines, and must actively participate in the various data management advisory groups.
- All available data sources should be checked for validity via accepted quality assurance and quality control measures, and once accepted, placed into an electronic format with a spatial data element for indexing in a relational database. The location and geometry of contamination plumes should be placed in a Geographic Information System (GIS) format.
- There is a large number of existing hard-copy water well drillers reports that need to be placed in a digital format and made accessible through the existing digital system.

Coordinate Research.
- The TGPC should form a research subcommittee to identify interagency research needs and to provide a coordinated approach for discussion with federal agencies for funding. The results of this work should be shared with the TCEQ for its consideration under the research model authorized under state law.

Increase Public Outreach.
- More water quality information is needed to develop assessments of water quality and health risk for the domestic/private well owner segment of the population.

- The state should undertake a voluntary program targeted toward private well owners, designed to identify problem areas and assist private well owners in understanding these groundwater quality issues.
- More support needs to be given to educational efforts for targeted geographic areas of concern for high concentrations of naturally occurring groundwater contaminants and on various treatment options available to the domestic/private well owner.
- Support is also needed for educational efforts to develop and deliver effective educational materials that target potential sources of contamination such as abandoned wells.
- Special effort should be made to develop educational programs designed to reach and serve the state's high-growth areas.
- The TGPC recommends that the state continue to support the efforts of the On-Site Wastewater Treatment Research Council, the Texas AgriLife Extension Service, the TCEQ's on-site wastewater program, and local governments in their efforts to develop and deliver effective educational material that addresses on-site sewage facility (OSSF) maintenance in order to prevent failures.
- Government agencies involved in OSSF regulation and outreach may want to consider developing programs specially designed to reach and serve the state's high-growth counties.
- The TGPC should establish links on its web site to key groundwater information residing at state agencies and educational institutions.

Commit to Development of Periodic Updates and Improvements to the State Groundwater Protection Strategy:

- The TGPC should review and update the *Strategy* every 6 years.
- The TGPC should conduct an analysis that will identify and rank threats to groundwater quality (taking into

consideration the vulnerability of groundwater resources and using available data), and prioritize possible actions that address those threats.

Strategy Recommendation:
The TGPC needs to strengthen the lines of communication and information sharing with the State's RWPGs. The lack of communication between these two groups is a gap in the TGPC's ability to coordinate the state's groundwater protection strategy with the state's water supply planning efforts led by the TWDB.

EPA, through the Clean Water Act, has provided grants to the state since 1985 to: (1) promote the coordination of groundwater protection activities of federal and federally-delegated regulatory programs; and (2) foster a more comprehensive approach to groundwater protection. In addition, starting in 1992, EPA has provided grants to the state under the FIFRA for groundwater protection activities specifically related to pesticide use and effects on groundwater.

The TGPC leads initiatives, in partnership with federal agencies, to develop a state groundwater protection strategy and implement PMP activities to protect groundwater from contamination. Current state and federal cooperative efforts include identifying potential improvements to the state's groundwater quality monitoring effort and ensuring that those efforts are consistent with national monitoring initiatives.

In addition, the TGPC regularly provides input at the national level to federal agencies through the Ground Water Protection Council (an association of state groundwater and underground injection control program directors), the State FIFRA Issues Research Evaluation Group (a group formed by state agricultural regulatory officials and EPA to discuss and evaluate pesticide matters affecting states), the National Water Quality Monitoring Council (an advisory group to the USGS and EPA), and other state and federal stakeholder and regulatory guidance groups.

The TGPC works closely with the USGS, the federal agency with hydrogeologic responsibilities that include national level geologic mapping and hydrologic studies. USGS participates in TGPC sponsored projects and subcommittees, providing both groundwater expertise and opportunities for state input into federally-sponsored research.

(Author's note)

There is little doubt that "protection" of groundwater from "contamination" is needed. Education on water matters, surface and groundwater, is important. Study of the state's aquifers must continue. Potential expansions of water resources are vital. All of these are needed and legitimate concerns.

But only government could claim that the taking of the property owners' underground water is "protecting groundwater".

A full study of the TGPC report leaves little doubt, once pass the carefully worded outlines, that the TGPC's Number One Objective is to provide as much cover as possible for the numerous local and state governmental agencies spread around the state, and their programs for control of groundwater.

The job of the TGPC is to justify all the current regulatory, outreach, research, and educational programs of all the agencies that have been given authority for controlling the state's groundwater resources.

A large component of their job includes working with these agencies to report areas where control over groundwater can be expanded – protected – either by the implementation of additional responsibilities or through the coordination of activities and policies.

The road map of the TGPC and the agencies they protect leads in one direction: Total control of groundwater.

"Protection," as it now stands, will allow the state's courts to rule that groundwater control is "for the public benefit" – and will result in a legal plunder of the landowner's historical absolute ownership of groundwater.

19 – THE PUBLIC BENEFIT

"When politicians talk about a public interest or a public good or the public benefit, they are expressing a philosophy of failure, the creed of ignorance, and the gospel of envy; its inherent long range or long term virtue is the equal sharing of misery." (unknown)

Government subsidies keep municipal water cheap.

Most cities today, and large water supply companies, depend upon surface lakes for their source of municipal water.

The federal government constructs the lakes using tax dollars. Water from the lakes is given free or sold at below the development costs to a sponsoring city. The costs of pipe lines to move the water to the cities, the water treatment facilities, and much of the local city's infrastructure to deliver the water to individual users are paid by government, either in direct payment or by grants.

All such costs are subsidies. Generally, the cost of the water sold to the individual users is not even close to the *actual cost* of the water.

Marketing experts agree that more realistic water pricing and improved water management will significantly cut water use.

But, the fact is, not much can be done to create water where water is needed.

One way or another, Texans will get the water they need, if not the water they want, perhaps not at the price they want.

City governments can find supplies—by desalinating water, recycling sewer water, capturing and filtering storm water from paved surfaces, etc. And among cities and industry, however, demand can be cheaply and quickly slashed — with a variety of conservation and efficiency measures, with higher rates for water wasters, and with better management policies.

It will not be politically popular, but it is the thing to do for those areas which need additional water but don't want to pay for it.

But none of these will create more water.

When it comes to water, nature did not always distribute assets fairly. Or perhaps, mankind has done a poor job of selecting places in which to live?

We realize that in some communities, the problem is often a lack of infrastructure — wells, pipes, pollution controls, and systems for disinfecting water.

Though politically challenging to execute, the solutions are fairly straightforward: Investment in appropriately scaled technology, better governance, community involvement, proper water pricing, and training water users to maintain their systems. Better management and efficiency will stretch the last drops.

We hear and read of regions facing scarcity because of over-pumped aquifers, of farmers switching from flood to drip irrigation; and communities concerned about replenishing aquifers. We also hear of global warming, climate change and a host of other dangers to survival.

But like global warming, much of what we read and hear assumes a different shape under the light or reason.

The fact is there has never been an aquifer go completely dry.

In states where water is controlled, farmers and ranchers are at the mercy of government propagandists, who point at them as the largest water users and the lowest rate-payers. (The key words in that sentence being "controlled" and "ratepayers".) They ignore the fact that "lowest rate-payers" do not apply to cities and water supply companies who enjoy subsidized water costs.

Those doing the controlling *always* want more money and the weakest politically must provide it. That's the law of the jungle: *Government 101.*

Still, the time is coming when some farmers and ranchers — the largest water users and the lowest ratepayers(?) — may find that the price of pumping water may exceed the price they receive for the product being watered. Drip irrigation, whether on the farm or in a large heavily populated residential subdivision is a good way to save water – and even a better way to save money on your water and energy bills.

Common-sense is a great motivator.

But cities which need additional water resources are not concerned with common-sense. They want additional water resources.

They can demand their politicians build more surface lakes to impound more water for their use. But not all areas of Texas have an abundant rainfall, and coastal areas must keep a fresh supply of water flowing into the Gulf. This means that capturing additional water must be done in some other area.

Often, too, building surface lakes is costly, time-consuming, and politically unpopular in the area where a lake is to be constructed. Usually, lakes are developed after a fight; a fight that the property owners within the lake-site loses.

Oh, lakes to impound water can be built and the impounded water can be moved vast distances to where it is needed – but it takes years of political in-fighting and many millions of dollars to do it.

The state already owns all the surface water, and can do anything it wants to do with the water.

So, Texas' water planners are enviously eyeing a water-rich area of NE Texas, as a location for several surface lakes. But what the planners really want is the regions' abundant underground water.

They have quietly suggested to cities needing water that control of underground water should be given to the state. Implied is the promise that the state would "take" the groundwater of the state's property owners, and give it to the cities and other large water users.

The government paid water planners point with alarm at areas that need water, the economic consequences of potential water shortages for business and industry, cattle dropping from thirst, crops dried up from a lack of water, and painful scenes of homes without water and "thirsty children".

Yes, what water planners see is a terrible future – without them.

And it is that future that water planners sell – without them.

Water planners view themselves – with pride – as society's saviors.

Of course, *if only YOU* had the knowledge and the wisdom on how to use your personal knowledge to save our countryside, our cities, our nation and the world, wouldn't you view yourself with pride, too? Regardless, of how foolish such conceit is?

But too many of us accept without question the conceit being sold to us from water planners.

Does conservation harm us? No.

Is stewardship of our assets, our families, our communities, towns and cities, and of the resources of our region, state, nation and the world needed? Yes.

But how much governmental control is really needed in our lives?

Why is government, made up of those much like us, better as a steward of our assets, families, communities, and resources than we are as landowners and neighbors?

Perhaps, we and our neighbors would not impose as much control as the water planners are offering. And they are not even offering to do it free; they want us to pay for it, to pay them for telling us what we must do and how we must do it, and when.

If we, as individual property owners, won't take care of those things which we own and which help provide us a living or have the potential to make our lives better, why would someone else do it better?

What makes a bureaucrat, a politician, or someone who wants what you have, better at taking care of what you own?

Oh, they point at "protecting you against your neighbor" and "the public benefit".

But are *all* your neighbors out to harm you? The water planners do not say *some might* harm you, but imply all are trying to do you harm.

But: Aren't the water planners someone's neighbor, too? And IF everyone's neighbors are all untrustworthy . . .

And property owners have allowed such illogic shape the debate.

But when did property owners become a separate group from "the public"? When were the property owners thrown out of the general society, and who did it, and why?

Even more importantly, why did we let them?

Perhaps, the most important question of all is:

"Why is *'the public benefit'* reserved only for those who want your property?"

PART THREE:

The Response of Landowner Organizations

&

Author's notes and observations

NOTE:

The following information concerns groundwater issues, questions about, and recommendations on ownership rights and regulations, as determined *by collaboration among several of the state's larger voluntary associations and organizations* that deal in matters of importance to a variety of Texas' property owners.

It is likely that many of these recommendations will be incorporated into Texas water law.

The information was gathered from a report prepared for use and dissemination by the sponsoring groups.

Comments at the end of this section, disagreeing or disapproving of the positions expressed by these organizations herein, are those of this author, and are included for comparison and study only.

Jake Street

Groundwater Recommendation by
Texas Landowner Organizations and Associations

Texas Groundwater Issues:
Ownership Rights and Regulation
Revised Oct.25, 2010

Acknowledgements:

These documents were prepared by Texas and Southwestern Cattle Raisers Association, Texas Wildlife Association and Texas Farm Bureau.

Many associations and organizations have joined this effort to protect property owners private property rights in groundwater. These documents reflect a joint effort, shared philosophy and commitment to groundwater ownership in place and local regulation as described herein. This is not a new association or organization and the initiative respects the autonomy of the individual supporting associations or organizations.

Please contact one of the (following) individuals or one of the participating associations or organizations for more information and how to participate in this effort.

For more information, please visit our website at www.groundwaterownership.com

Texas and Southwestern Cattle Raisers Association
1-800-242-7820

Association of Texas Soil and
Water Conservation Districts
    512-818-1660

Texas Wildlife Association
1- 800-839-9453

Exotic Wildlife Association
830-367-7761

Riverside and Landowners Protection Coalition

South Texan's Property Rights Association
361-348-3020

Texas Association of Dairymen
866-770-4823

Texas Feeders Association
806-358-3681

Texas Land and Mineral Owners Association
512-479-5000

Texas Forestry Association
936-632-8733

Texas Poultry Federation
512-248-0600

Texas Farm Bureau Austin Office
512-472-8288

Texas Sheep and Goat Raisers Association
325-655-7388

Introduction

According to estimates by the Texas Water Development Board (TWDB), by 2060 Texas" population will more than double, and its water demand will increase by 27%. Because groundwater from Texas' aquifers supplies over half the state's water, it is imperative that groundwater resources be managed to provide for current and future use.

In 1997, the Texas Legislature made significant changes to water planning and groundwater conservation district management. The new Regional Water Planning process highlighted the future water needs as well as where the water would be procured to meet those needs.

This process stressed the need for many areas without groundwater regulation to create groundwater conservation districts to protect this finite natural resource.

In 1999, only 45 groundwater conservation districts existed. These districts were primarily located west of the IH-35 corridor. Today, 96 groundwater conservation districts exist covering 144 counties. Of the 10 million acre-feet currently produced in Texas, 9 million acre-feet of this production is within a groundwater conservation district. Therefore, over 90% of the groundwater produced today could be impacted by the regulation of groundwater conservation districts, Edwards Aquifer Authority, or the Harris/Galveston.

Subsidence Districts.

Groundwater is and always has been an integral part of the land, and secure, protectable rights assure conservation and stewardship of groundwater [see white paper titled "Groundwater Ownership and Regulation in Texas"]. Texas and Southwestern Cattle Raisers Association, Texas Wildlife Association, and Texas Farm Bureau have joined forces to initiate a growing number of associations and organizations protecting property owners' private property rights in groundwater, while supporting reasonable, science-based regulation for the long-term sustainability of groundwater resources. To advocate these ideals, the supporting associations and organizations are establishing an educational program to assist members, property owners, legislators, and policy makers in understanding current groundwater ownership and regulatory issues in Texas.

Joint Position Statement on Groundwater Ownership

Groundwater is an integral part of the land and is owned by private landowners. The Texas Constitution and more than 100 years of case law supports this position. Secure, protectable property rights best assure conservation and stewardship of all resources, including groundwater.

As the demand for groundwater in Texas increases, it is important that groundwater continues to be recognized and reaffirmed as vested, real property of private landowners. Private landowners and their productive open land are keys to an effectively functioning water cycle. Their active and informed stewardship of land and water resources benefits all Texans.

Like other private property in Texas, groundwater is subject to reasonable regulation. This ensures that private landowners are treated fairly (afforded due process), property rights are respected, and that all private landowners maintain the ability to use groundwater for any beneficial use.

Just as it makes sense for school districts to be governed by local citizens, it makes sense for groundwater to be governed by local citizens, which is why we support local groundwater conservation districts. It is better and more effective for private landowners to work with their neighbors, rather than a distant state agency.

However, for groundwater conservation districts to function as they were intended, recognition and reaffirmation of groundwater ownership is needed so groundwater conservation districts are consistent in this interpretation across the state. All groundwater conservation districts must recognize that groundwater is the property of private landowners and use sound scientific principles to develop reasonable regulations that ultimately will ensure the beneficial use of groundwater.

We support protecting and reaffirming that groundwater is the vested, real property of private landowners for the following reasons:

1. State law is clear that groundwater is the vested, real property of private landowners, but some continue to challenge the law. Private landowners must defend and reaffirm their ownership of this property and all constitutionally mandated private property rights in the regulatory, legal, and legislative arenas to protect the resource for the benefit of all.

2. Private landowner ownership of groundwater encourages good stewardship and promotes accountability. The way private landowners, acting as land stewards, manage

their property directly influences quantity and quality of groundwater available to all Texans. Vested ownership with local control also equitably balances conservation and use.

3. Private landowner ownership of groundwater provides more certainty and balance in water planning. With groundwater ownership reaffirmed, water planners can concentrate on how best to use groundwater to meet the state's critical needs instead of arguing about who owns it. This helps balance rural water-producing areas and urban water-consuming areas, without jeopardizing potential growth in any area of the state. Recognition of all landowners' rights ensures that the value of available groundwater resources is shared by all property owners, not just a select few.

Legal Review and Facts

Groundwater Ownership in Texas
1. *Absolute Ownership Doctrine*
Groundwater ownership in Texas begins with the *absolute ownership doctrine*, which establishes a property owner's vested right in groundwater in place below his land—a vested "real property" right that provides ownership in place. The Texas Supreme Court adopted the absolute ownership doctrine in the Houston & Tex. Cent. Ry. Co. v. East case of 1904: An owner of soil may divert percolating water, consume or cut it off, with impunity. It is the same as land, and cannot be distinguished in law from land. *So the owner of land is the absolute owner of the soil and of percolating water, which is a part of, and not different from, the soil.*

Texas courts have consistently applied this doctrine to groundwater ever since.

A corollary or subset of the absolute ownership doctrine is the *rule of capture*, which defines a property owner's "liability protection" for producing groundwater pursuant to his vested real property right. The rule of capture allows a property owner to produce groundwater for any beneficial purpose without liability to his neighbors. *The rule explains the manner in which a landowner may exercise his property rights in groundwater, not whether the property rights exist.*

2. *Right to Groundwater in Possession*

In addition to the real property right associated with absolute ownership, property owners have a vested right to groundwater that is legally produced and reduced to their possession. Once water is withdrawn from its underground source, it becomes "personal property" subject to sale, commerce, and taxation. This vested personal property right is different from the vested real property right associated with the absolute ownership of groundwater in place.

3. *Arguments about the Ownership Right*

Some argue there is no vested right in groundwater in place because, under the rule of capture, landowners have no protection from drainage by neighboring wells. They argue property owners only have a right to groundwater that does not vest until it is produced and reduced to possession. This argument is flawed.

First, there have always been limitations on the rule of capture. In *East* the Court acknowledged the rule of capture could be modified by "express contract and positive authorized legislation." Since *East* the Court has identified additional exceptions that provide protection to adversely affected landowners, such as willful waste and negligently caused subsidence.

Second, the rule of capture has never been interpreted to divest a property owner of ownership in place. Rather, throughout its history—both its oil and gas and its groundwater history—the absolute ownership doctrine and its corollary, the rule of capture, have been premised on ownership in place, which vests with ownership of the surface estate.

Third, while the Texas Supreme Court has sometimes criticized the rule of capture, it has never considered or recognized an alternative theory that would undermine the principle of absolute ownership in place. For instance, in the 1999 *Sipriano* case, the Court declined to repeal the rule of capture in favor of the American (reasonable use) rule. *Both of these liability rules are based on the principal of ownership in place. Thus, even if the Court had made a shift from the rule of capture to the American rule, it would not have changed the underlying ownership interest such rules seek to protect.* This is confirmed by the Court's cautionary statement that "any modification of the common law [rule of capture] would have to be guided and constrained by constitutional and statutory considerations."

The Takings Clause Limits Regulation of Groundwater Rights

Because property owners have a vested real property right in groundwater in place, the state may not unreasonably limit a property owner's groundwater rights without compensation. Given this constitutional limitation, the Legislature has expressly recognized and preserved landowners' ownership interests in groundwater in section 36.002 of the Water Code: *The ownership and rights of the owners of the land and their lessees and assigns in groundwater are hereby recognized, and nothing in this code shall be construed as depriving or divesting the owners or their lessees and assigns of the ownership or rights, except as those rights may be limited or altered by rules promulgated by a district.*

Therefore, while groundwater regulation may limit the exercise of groundwater rights, the statutory and common law still confirm the principle of ownership in place. Any groundwater regulation must reasonably balance future needs with property owners' private property rights.

The clash between regulation (e.g., the DFC process) and property owners' ownership rights is coming to a head at a time when the Texas Supreme Court has once again been asked to consider the nature of a landowner's ownership interest in groundwater. *Edwards Aquifer Authority v. Day*, 274 S.W.3d. 742 (Tex. App.—San Antonio, pet. pending). The Attorney General and the Edwards Aquifer Authority have argued that property owners' groundwater rights in place are not vested, and thus regulation by GCDs can never constitute a taking. If groundwater rights are not protected constitutionally, there is no recourse for a property owner when that property right is taken or restricted. The threat to property owners statewide is apparent: Unreasonable and restrictive decisions by local GCDs can profoundly limit landowners" private property rights. Limiting these rights could deprive all Texas citizens of the economic benefits of the resource.

Unreasonable Groundwater Regulation Will Hamper Economic Development

If Texas cannot meet its water needs, Texas" economy will suffer. According to TWDB, "without water, farmers cannot irrigate, refineries cannot produce gasoline, and paper mills cannot make paper." If future water needs are not met it could cost businesses and workers in the state approximately $9.1 billion per year by 2010 and $98.4 billion per year by 2060. State government could lose $466 million in tax revenue in 2010 and up to $5.4 billion by 2060 due to decreased business activity as a direct result of insufficient supply.

In a 2009 report, *Liquid Assets: The State of Texas' Water Resources*, Comptroller Combs cautioned that policy makers and GCDs must consider the impact their actions could have on a property owner's private property rights in groundwater, including the potential economic impact. As she poignantly stated: "[R]ecent court rulings have affirmed the state's long-held position on ownership of private property, as codified in . . . the Private Real Property Rights Act . . . . Groundwater is the property of the owner of the land overlying the aquifer, and efforts to interfere with this right could result in both uncertainty of ownership and enormous economic consequences for our state."

Recognizing Property Rights with Science-Based Regulation Is the Best Approach

The best way to manage and prepare for the state's future water needs is reasonable, science-based regulation that continues to recognize vested real property rights in groundwater. Ownership of groundwater in place provides for the continued development of a groundwater market that can meet the state's critical needs within regulatory limits. First, private ownership of groundwater provides certainty, consistency, and balance in water planning. With a consistent, protectable property right, efficiency and economy can flourish.

Second, ownership in place encourages good stewardship and promotes accountability. Ownership in a market based system will encourage and incentivize property owners to closely analyze the quantity and quality of their groundwater. A landowner is better positioned to ensure a clean and ample supply on his property than is a distant state agency. Finally, ownership in place is critical to preventing development of an artificial groundwater market. True market demand will establish the pace and price of groundwater development, which will depend on factors such as the distance to and reliability and sustainability of supply. It will promote a balance between rural water-producing areas and urban water-consuming areas, without jeopardizing potential growth in any area of the state. A groundwater market, based on vested private ownership and sound science, allows economic development to occur but at the same time allows property owners to continue their existing use should they so choose. The alternative is loss of rights without remedy, and the state deciding who can use water and for what purpose.

Frequently Asked Questions

1. Why was this effort initiated?
Property owners' ownership of groundwater is under attack in the courts and in some groundwater conservation districts (GCDs). The supporting associations and organizations believe private ownership, coupled with reasonable, science-based regulation, is the best way to manage Texas" groundwater resources and assure stewardship and conservation. A balance can be struck between property rights and reasonable regulation while assuring protection for public and private investments and water needs in every part of the state.

The Attorney General has urged the Texas Supreme Court to not recognize ownership of groundwater in place. Some groundwater conservation districts (GCDs) are adopting rules and pumping limits that will deprive property owners of the right to produce groundwater in the future. Given these trends, the supporting associations and organizations believe its effort is necessary to ensure that groundwater continues to be recognized as a vested private property right. Private ownership fosters conservation and ensures the future beneficial use of groundwater. Reasonable regulation ensures that all property owners are treated fairly and that property rights are respected. Private ownership coupled with reasonable regulation is the best method to protect property rights and manage groundwater resources for the benefit of all Texans. Vested ownership incentivizes property owners to maintain and manage critical open space lands that contribute to and form the geological sponges that are physically integral to the land and a connected component of groundwater resources.

For instance, if groundwater rights are not vested until reduced to possession, then the only way to protect them is to produce the groundwater before it is too late. This is contrary to protecting and conserving the resource, and is bad public policy. With vested ownership of groundwater in place there can be an appropriate balance. Recognition of all property owners' rights ensures that the value of available groundwater resources is shared by all property owners, not just a select few. Reasonable regulation should recognize ownership interests and existing use, while allowing water planners to manage the resource for long-term sustainability and assure long-term public benefits. This type of balance is critical for economic development in the state.

2. *How does the public benefit from reaffirming that groundwater rights are vested?*

It supports conservation of Texas' groundwater resources, recognizes the private property rights of Texans, fosters reasonable and fair regulation that encourages consistent treatment, promotes accountability for management and conservation, and allows a regulated groundwater market to develop as society demands it, and in a manner that includes all property owners.

The supporting associations and organizations support the preservation of property owners" vested property rights in groundwater as critically important to the management and conservation of groundwater resources in Texas. Private ownership of groundwater encourages and incentivizes good stewardship and promotes accountability. As the demand for groundwater in Texas increases, private ownership in a market based system will encourage close analysis of the quantity and quality of a property owner's groundwater.

This system already exists today in many areas of the state. A property owner in coordination with the local GCD is better positioned to ensure the quality and supply of groundwater on his property than is a distant state agency.

Private ownership of groundwater provides certainty, consistency, and balance in water planning, allowing planners to concentrate on how best to use groundwater to meet Texas' critical needs. It affords the ability to achieve a balance between rural water-producing areas and urban water-consuming areas, without jeopardizing potential growth in any area of the state. Ultimately, private ownership of groundwater supports free enterprise, allowing the continued, reasonable development of a regulated market for groundwater that includes all property owners. The regulation of this market is governed by limits established by the Legislature. If groundwater rights are not vested, they can be unreasonably limited or eliminated, leaving the property owner with no right to produce groundwater from his property and no remedy for its loss. This could potentially affect land values, lending institutions, financing of long-term existing and planned public and private water projects, and the Texas economy.

Ownership of real property fosters conservation because it is an integral part of the land and is tied to management activities affecting land conservation and stewardship. The suggestion by some that you don't own groundwater until you produce it is a serious disincentive to conservation, and a policy that encourages unnecessary production.

3. *As a property owner, why should I care or get involved?*

The other option is to let the government take your property without a fight. The Texas Alliance of Groundwater Districts, the Edwards Aquifer Authority, and even the Attorney General, are arguing in court that property owners do not have a constitutionally protected property right in groundwater.

Groundwater is an integral part of the land. Texas law has repeatedly affirmed the private ownership of groundwater for over 100 years, although the courts have not addressed the extent of its constitutional protection. However, as Texas" population grows and water becomes scarcer,

landowners" groundwater rights are increasingly under attack. The supporting associations and organizations support the constitutional position that the law has always established groundwater ownership as a vested property right. On the other hand, the State and some GCDs are attempting to redefine groundwater rights as unvested property rights. For example, the Texas Alliance of Groundwater Districts, the Edwards Aquifer Authority, the Attorney General, and others filed briefs with the Texas Supreme Court arguing that the property owner does not own groundwater in place. Such a characterization is critically important to a property owner's ability to exercise groundwater rights. It would mean your groundwater rights could be taken (curtailed or halted) without any recourse or compensation. A groundwater system that unreasonably and unfairly restricts use of your property affects the value of that property, and your ability to conserve, produce, use, or sell that property. Even property owners" exempt use and historic use of groundwater would be at risk. Moreover, those who have historically conserved the resource (by producing less or not producing at all) could be left with few or no rights at all. Ultimately, an ownership right that is not vested is less valuable and much harder to protect.

4. *What is the current law regarding groundwater ownership?*

A property owner is the absolute owner of the groundwater in place beneath his property, subject to reasonable regulation.

Private ownership of groundwater derives from the common law, beginning with the Texas Supreme Court's adoption of the *absolute ownership* doctrine in a case called *Houston & Texas Central Railway Co. v. East*: "An owner of soil may divert percolating water, consume or cut it off, with impugnity (sic). It is the same as land, and cannot be distinguished in law from land. So the owner of land is the absolute owner of the soil and of percolating water, which is a part of, and not different from, the soil." Tex. 146, 150, 81 S.W. 279, 281 (1904).

The Texas Supreme Court has reaffirmed this principle numerous times since 1904. From a statutory perspective, the Texas Legislature has also recognized a property owner's ownership interest in groundwater in Chapter 36 of the Texas Water Code. Section 36.002 says, "The *ownership and rights* of the owners of the land and their lessees and assigns *in groundwater* are hereby recognized, and nothing in this code shall be construed as depriving or divesting the owners or their lessees and assigns of the ownership or rights, except as those rights may be limited or altered by rules promulgated by a district." As private property, groundwater rights are also protected by the constitutional mandates that private property cannot be taken without compensation and due process of law.

5. *How can you own groundwater if you can't quantify how much is below your property, or if it moves from one parcel to the next?*

Property owners have a *real property right* in the groundwater in place below their land, but they do not own a particular molecule of groundwater when it is below their land. Property owners have the right to produce groundwater from their land, subject to reasonable governmental regulation and other existing common law restrictions such as for beneficial use and prevention of waste. Once groundwater is produced and reduced to possession by the property owner, the groundwater becomes *personal property*, subject to sale and commerce like a cow or tractor. Groundwater is similar to oil and gas in this respect: it usually is not stationary and some property owners have more of it below their land than others. But this does not mean landowners don't have a real property right in oil and gas in place below their land.

6. *How can you own groundwater if you can't protect it from your neighbor or others?*

The rule of capture is simply the rule of liability associated with the exercise of groundwater rights under the absolute ownership doctrine. The rule of capture addresses how the property right may be exercised, not whether it exists.

Property rights opponents argue that an essential element of ownership is the right to exclude others, and that groundwater cannot be owned in place because the rule of capture precludes the surface owner from preventing drainage of his property by others. This argument is belied by nearly 100 years of oil and gas law recognizing the concept of ownership in place alongside the rule of capture. The Texas Supreme Court has consistently recognized both the rule of capture and ownership in place for groundwater. Moreover, you do have some protection from your neighbors under the rule of capture, such as for wasteful use and negligent subsidence. The rule of capture is also subject to reasonable regulation. Absolute ownership in place and the rule of capture are not contradictory legal principles, and recognition of ownership in place will not negate the ability of the state to regulate groundwater.

7. Do you support mandating a correlative rights approach to groundwater management?

No. The ownership of groundwater in place can be recognized without mandating every district use a correlative rights system to allocate groundwater.

Supporting associations and organizations, however, support every property owner's right to be given the opportunity to produce a reasonable amount of groundwater. It is the responsibility of each GCD to determine what management and regulatory practice best suits their local situation.

8. Are you proposing to cut off existing users, and asking GCDs to cancel and re-issue new permits?

No. There should be no need to cancel or re-issue permits unless those permits were issued under a system that illegally curtailed or prohibited use by others. However, existing users may have their pumping reduced in the future so that other property owners can exercise their rights to the groundwater.

A vested right to the groundwater in place strengthens existing permitted rights. Without vested ownership in place, the only right to groundwater an existing user has is those granted by the GCD"s rules. Therefore, the rights are wholly vested in rules of the district, and can be taken away with a change in the district's rules. Also, such rights do not necessarily transfer with the property because the district's rules may require a new property owner to reapply for a permit. To the contrary, with ownership in place, the right to produce groundwater remains with the land or property owner and cannot be taken away by a change in GCD rules. However, existing and historic users must recognize that most of them made investments through groundwater under rule of capture where all property owners had an equal right to capture groundwater and there was no protection of the investment. If a GCD's rules now provide protection for existing use, it is a new protection that did not exist when the investment was made.

After sound science identifies the extent of groundwater resources in a district, and only if all permit holders will "benefit" from reductions, would district-wide reductions make sense. Even then, an existing user could make up for a loss on investment through conservation measures or the market, by selling or purchasing water rights.

Even when groundwater resources are severely restricted, market pressures will dictate who desires to continue an existing use, begin a new use, or sell to someone else for another use. No property owner would be denied groundwater; the landowner may simply have to purchase additional rights if the district has to reduce production by existing users.

9. Are you suggesting that GCDs cannot say "no"? Won't reaffirming private ownership deplete aquifers because GCDs cannot say "no" to permit applicants?

Yes. The supporting associations and organizations believe ALL property owners have a vested right to capture groundwater on their property. However, we are not suggesting that GCDs cannot regulate groundwater production.

GCDs have a statutory responsibility to protect groundwater resources. This would be impossible if the district could not limit a property owner's right to capture groundwater. However, the GCDs should not say "no" to a property owner with a valid, beneficial use. The GCDs should treat all property owners fairly. If there is a situation where a permit should be denied or severely limited, the GCDs must evaluate whether that decision would constitute a regulatory taking. There could be situations where denying a permit may not be considered a taking by the courts. It will depend on each individual situation.

10. State law seems clear, so why reaffirm the obvious? What is happening at GCDs or in the courts that you don't like?

With the desired future condition process well underway and likely to result in caps on production, and the Texas Supreme Court addressing whether groundwater rights in place are constitutionally protected, a perfect drought could be brewing for Texas property owners.

Local GCDs, at the direction of the Legislature, are required to adopt "desired future conditions" (DFCs) for the groundwater resources in each Groundwater Management Area. These DFCs will be used to calculate "managed available groundwater," which will be used by districts to limit the amount of groundwater that can be pumped in each district. In most instances, this will result in a cap or limit on production. GCDs must set production limits to achieve the desired future conditions of the Groundwater Management Area that will affect water rights allocations among users. Some GCDs believe they can say "no" to new permits to pump groundwater because they believe there is no vested property right in groundwater until it is produced, and thus no regulatory taking for denying a property owner the right to produce. Under this same logic, some GCDs will adopt rules restricting even historic uses.

The supporting associations and organizations seek to ensure that GCDs allocate rights to produce groundwater that will respect private property rights.

In the courts, the Texas Supreme Court is currently considering the appeal of a major groundwater rights case: *Edwards Aquifer Authority v. Day*, 274 S.W.3d. 742 (Tex. App. — San Antonio 2008, pet. pending). This case squarely presents whether governmental regulation of groundwater can amount to a taking of vested private property requiring compensation. This is an important case because, while the Texas Supreme Court has repeatedly reaffirmed the absolute ownership doctrine of groundwater, it has not done so without criticizing the rule of capture aspect of the doctrine and suggesting that another system may be needed in the future.

In the *Day* case, the Edwards Aquifer Authority, the Texas Alliance of Groundwater Districts, and the Attorney General have argued the absolute ownership doctrine does not vest any ownership interest in groundwater until that groundwater is actually produced and reduced to possession. Under this theory, GCDs and the legislature could unreasonably regulate groundwater — in fact, completely halt its production — without any liability for depriving a property owner of that property right. If adopted, this position would completely undermine the private property rights of Texas property owners.

11. Didn't recent court cases, like City of Del Rio and Guitar, resolve this issue?

Previous court cases have not answered whether regulation of groundwater production can constitute a taking of private property as applied to a particular property owner's situation, which issue is squarely before the Texas Supreme Court in the *Edwards Aquifer Authority v. Day* case
.

Prior to *Day*, the Texas Supreme Court was most recently presented with the takings issue in a 1996 case called *Barshop v. Medina County Groundwater Conservation District*, but avoided providing a complete answer. In more recent cases, such as *City of Del Rio* and *Guitar*, the issue has not been squarely presented.

In *City of Del Rio v. Clayton Sam Colt Hamilton Trust*, the San Antonio Court of Appeals did address the issue of groundwater ownership. While the court reaffirmed the absolute ownership theory, and recognized the ownership interest in groundwater, the court was not presented with the question of how far governmental regulation can go before a regulatory taking of groundwater rights occurs. And, since the Texas Supreme Court did not agree to hear the case, the decision of the appeals court does not have statewide effect.

In *Guitar Holding Co. v. Hudspeth County Underground Water Conserv. Dist. No. 1*, 263 S.W.3d 910 (Tex. 2008), the Texas Supreme Court voided the rules of a groundwater conservation district that granted certain historic irrigation users the perpetual (and exclusive) right to transfer and sell their groundwater outside the district for an entirely new use, while all other new users were effectively barred from producing groundwater. While *Guitar* does establish some limits on a groundwater district's authority, the case did not address whether the district's rules constituted a taking of private property.

12. Why is legislation necessary?

The Texas Water Code does not clearly state that the rights of property owners to groundwater is a vested right to groundwater in place.

Section 36.002 of the Water Code recognizes property owners' private property rights in groundwater, but qualifies that those rights may be "limited or altered by rules promulgated by a district." The supporting associations and organizations believe the context of this limitation relates to enacting reasonable and fair regulation, not undermining constitutional property rights protections that guarantee ownership. Thus, while the Supreme Court could provide some baseline protection to property owners, Chapter 36 of the Water Code must be amended to ensure that GCDs must recognize landowner property rights when adopting rules and issuing permits. If the law is not clear, it will provide an opportunity for some districts to attempt to deny property owners their rights to groundwater.

13. Does reaffirming vested ownership of groundwater in place change any existing major groundwater issues facing the legislature and GCDs?

It does not change the issues, although it may change how the issues are resolved. The legislature and GCDs are still facing a growing population and a limited resource. They still must find a reasonable way to manage the resource for the future. Regardless of how the ownership interest is defined, the legislature and GCDs still must find a reasonable way to protect and conserve our state's groundwater resources. There are major issues facing groundwater policy makers in this effort. For instance, in the context of planning for long-term management, policy makers must address how to share the resource among owners and users of groundwater. Should it be according to the rule of capture, equitable distribution, prorata sharing, reasonable and prudent use, or some combination of these? They must deal with existing uses, and the balance between agricultural irrigation use versus urban and commercial use. They must deal with the sale and transfer of water rights outside of a groundwater district or an aquifer. They must deal with existing contracts governing groundwater production, and potential property fragmentation if groundwater were to be no longer privately owned. Policy makers must deal with all these issues in the framework of sound science, which is only now being developed. For instance, they must understand how much groundwater is available in an aquifer, what are the recharge rates, and how much can be produced.

Many of these issues will continue to exist and need to be addressed even if vested ownership is reaffirmed. However, vested ownership in groundwater ensures that any regulation seeking to achieve these policy goals must be reasonable. It ensures that existing and historic uses will be protected. It also solidifies the already existing and extensive private contract market for groundwater, which is subject to reasonable regulation and beneficial use, as has always been the case. Reaffirmation of vested ownership in groundwater is the only way to create a true water market that appropriately balances the varying interests across the state. If vested ownership no longer exists, certain options will be removed from the table, and any market would be artificial and subject to political maneuvering.

14. Are the supporting associations and organizations opposed to GCDs?

Not at all. A GCD that recognizes in its rules the ownership interests of property owners, acts within its authority, and uses sound science to guide its rulemaking, provides for effective regulation of groundwater resources.

Local GCDs are the best method for regulating and conserving a local resource. These districts, in consultation with their local constituents, are in a far better position to regulate their local resource than is a distant state agency. But districts must act within constitutional limits and their legislative authority. They must also act pursuant to sound scientific principles regarding groundwater availability and establishing desired future conditions.

15. What will happen to the value of my land if vested ownership of groundwater in place is reaffirmed?

Vested ownership in place will protect existing values and may even enhance the values of some properties.

Subject to proper regulation and certainty in groundwater production, property values could improve. On the other hand, an overly restrictive regulatory system and uncertainty in groundwater production would adversely affect property values. From a lending perspective, property subject to reasonable regulation and certainty of use is more favorable than property subject to problem regulation and uncertainty of use. Recognition of all property owners" rights ensures that the value of available groundwater resources is shared by all property owners in a community, not just a select few.

For existing users with existing permits, recognizing the rights of all property owners will not affect the current ability to capture groundwater under an existing permit. Moreover, GCDs have the ability to provide greater protection to historic and existing uses.

16. If I have vested ownership of groundwater in place, can it be valued and taxed?

Groundwater in place cannot be taxed under current law.

Groundwater is an integral part of the surface estate, just like limestone, sand, and gravel. The supporting associations and organizations maintain that property owners have a right, and have always had a right, to the groundwater on their property, just like they have a right to mine or quarry limestone, sand, or gravel. These "substances" cannot be taxed unless they are being extracted and sold. Otherwise, they are all taxed "in place" as part of the market value of the land. If the law clarifies that property owners have a vested right in groundwater in place as part of the surface estate, just like all the other substances of the surface estate, that does not change how the substances of the surface estate are taxed.

The Texas Supreme Court addressed these same issues with respect to limestone in *Gifford-Hill, Co. v. Wise County Appraisal District* in 1992. In that opinion the Court clarified that property can only be taxed if the Texas Property Tax Code authorizes it to be taxed. The Texas Tax Code does not authorize the taxation of groundwater. Therefore, if the Texas Supreme Court or the Legislature establishes that groundwater is the vested property of the property owner, it still cannot be taxed unless legislation is passed amending the Tax Code to authorize such a tax.

17. If a property owner sells his groundwater rights, how will that affect use of the land in the future? Can future owners of that land still get access under the domestic and livestock exemptions even though someone may have already sold all the groundwater rights?

A property owner is generally free to contract as he desires. Thus, future groundwater use depends on the conveyance documents used to accomplish the sale or transfer.

If a property owner wants to reserve all water rights or just a portion, he may do so. If a property owner wants to sell all water rights, but reserve access to groundwater for exempt uses, he may do so. Property owners should always consider whether and how to sever and reserve or sell water rights, and evaluate the terms of the conveyance documents carefully, and we always recommend advice of a competent, independent attorney before finally signing any documents. Of course, any production will have to be in compliance with the GCD's rules, no matter who owns the water rights.

We believe a reservation for exempt uses is good policy and should be required in all conveyances severing groundwater rights from the surface estate.

18. How does vested ownership affect groundwater marketing? Doesn't it encourage or fast-track such a market?

A groundwater market already exists in many areas of the state. It is relied upon by property owners and cities statewide. Vested property rights in groundwater in place do not by themselves create or fast track a groundwater market. The groundwater property right only directs how that market will develop subject to regulation by GCD rules and sound science. Recognition of all property owners' rights ensures that the value of available groundwater resources is shared by the community, not just a select few. Vested ownership in groundwater in place is critical to recognizing existing transactions and the development of future groundwater production, rather than an artificial market created by regulatory controls that harm land stewardship and conservation. Vested rights in groundwater in place only ensure that a true water market can exist, not whether or how quickly it will develop. Society and the economy will establish the pace and price for groundwater markets, which will depend on factors such as the distance to and reliability and sustainability of groundwater supply. Given these factors, markets for transfer of water are most likely to develop first, and perhaps only, in locations with abundant groundwater supplies near high demand, such as a major city.

Vested ownership does not prevent water marketing, as some contend. Oil and gas is a good example of how this would work.

19. Isn't a groundwater market bad? For instance, won't water go to the highest bidder?

A market already exists, and nothing done on the groundwater ownership front will change that. For example, in the Texas panhandle groundwater rights have been bought and sold with no detriment to the existing communities and their economies.

If a market develops in your area, you can choose whether to participate, and the local community through the GCD can determine if the groundwater, based on sound science, can be used for new uses. You may choose to conserve your groundwater, or continue its existing use. Or there may be an economic advantage for you to sell your groundwater rights. In all instances, you have the choice, and your existing groundwater use will be protected as a vested property right. A groundwater market, based on vested ownership rights in place and reasonable, science-based regulation, allows economic development to occur but at the same time allows property owners to continue their existing use should they so chose. Of course, there will be an economic incentive for the most cost effective use of the groundwater. Without a market, there are fewer choices, and there is little economic value to your groundwater resource outside of its use on your property.

20. How does groundwater impact the Texas economy?

Groundwater plays a critical role in the functioning of the state economy. In her letter brief to the Supreme Court of Texas, the Texas Comptroller of Public Accounts indicated the following: "Groundwater and the rights of landowners associated therewith play a fundamental role in the strength of the Texas economy. To cite but one example, agriculture, Texas' second largest resource-based industry, has an economic impact on the Texas economy of approximately $100 billion.' The vast majority of our state's total land area, almost 80 percent, involves some type of agricultural production.' Critical to the continued strength of agriculture in our economy is groundwater and it is the source of 73 percent of the water used in Texas' irrigated agriculture industry. In addition, communities across the state have invested a great deal of time and money in acquiring groundwater rights from landowners to support the water supply needs of their vibrant and growing economies. The continued sustainability and development of these water projects are important to our state's economic success. Texas has a substantial interest in protecting its economy by ensuring continued acknowledgment of established groundwater rights."

21. Will there be a plague of takings liability? Will it bankrupt my GCD?

No. A regulatory takings case is difficult to make against a GCD or any other governmental entity.

While some make this argument, the truth is a takings case is difficult to make. A property owner must meet a very high bar to establish a takings claim. Only if regulation is truly unreasonable, or if a property owner's rights are significantly restricted, or the property owner has made significant investment based upon future groundwater production, could a takings claim be successful. For instance, since the Edwards Aquifer Authority legislation was confirmed in 1996, the

Edwards Aquifer Authority itself has only identified three takings cases resulting from the sweeping limitations imposed by the Edwards Aquifer Authority Act. Lawyers will not be interested in such cases unless the facts are sufficient to meet the high bar. Moreover, property owners will be deterred from bringing frivolous or weak lawsuits because GCDs have the ability under Chapter 36 to recover their attorneys' fees from the property owner, if the district prevails in the suit. This forces attorneys and property owners to seriously evaluate their case.

22. If a taking is shown, who will pay?

The regulatory entity will pay. If the entity does not have adequate monetary resources, it may have to increase fees or taxes to compensate the property owner whose groundwater rights were illegally taken to the benefit of permitted users in the district.

Some argue that the compensation paid to these property owners may be at a significant cost to the local people or the state. While this may indeed be an unfortunate result of taking the rights of a property owners in some cases, it does not justify damaging a property owner without compensation. All property owners have a constitutional right to be compensated when government takes damages or, destroys their property or property rights.

23. What will the supporting associations and organizations do next?

The supporting associations and organizations have planned an educational and legislative program, to ensure that our members, our fellow Texas property owners, and our legislators are properly informed about this significant issue, and will continue to advocate for this issue.

Author's Opinion/Comments

We, the editorial "we", understand and sympathize with the position taken by those who live in groundwater control districts.

They have little choice on how they fight the war.

Once a groundwater district is established, it cannot, by law, be disbanded or dissolved or otherwise disposed of…it it there forever. Or so the water planners claim. Therefore, those in a district must fight for whatever victories they may be able to achieve.

Most statewide organizations' members live within existing groundwater control districts. Since the law doesn't allow districts to disband, the next best thing is to bring "control districts" under control.

And this surely is the long-term objective of those living within a district.

What possible political good could these state associations and organizations achieve by fighting the districts already imposed on their members, who have no current, legal way to get out from under the districts?

Their best course of political action is to take the position as stated in their publication, Ownership Rights and Issues: *"…A GCD that recognizes in its rules the ownership interests of property owners, acts within its authority, and uses sound science to guide its rulemaking, provides for effective regulation of groundwater resources.*

Local GCDs are the best method for regulating and conserving a local resource. These districts, in consultation with their local constituents, are in a far better position to regulate their local resource than is a distant state agency. But districts must act within constitutional limits and their legislative authority. They must also act pursuant to sound scientific principles regarding groundwater availability and establishing desired future conditions."

The political reality is that such an – *seemingly* - endorsement will garner endorsement from many districts governed by a *"board of disenchantment"*. It is likely this will be the basic philosophy that will be echoed by members of the Legislature responding to demands of property owners.

The first step in regaining ownership control is to take the first step.

And when your approval of GCDs consists of *"recognizes in its rules...ownership...act within constitutional limits...act pursuant to sound scientific principles..."* it's not exactly a ringing endorsement of a water district.

To an alert reader, it could sound as more of a warning.

Regardless of the perception, the fact is that the danger to all areas, including existing districts, is the sixteen Groundwater Management Areas authorized in Senate Bill 2, and strengthened under House Bill 1763.

These GMAs are designed to become super districts, covering all of Texas. Existing districts would be under the control of one of the regional GMAs.

Senate Bill 2, in 2001, passed at the urging of state water planners, established 16-GMAs and 16-Regional Water Planning Districts. It also authorized the GMAs to set more restrictive limitations on future groundwater use.

In 2005, House Bill 1763 increased the authority of the GMAs, giving them the power to set "Desired Future Conditions (DFC)" of groundwater resources in all of the state's aquifers. This DFC power allows GMAs to set a cap on overall production, deny permits, create special permit systems, exempt certain users, reduce current production of existing users, etc.

GMAs are a unit of state government and, as such, have taxation powers.

The water planners, through these GMAs, have set the stage for control of an area's groundwater whether it is or is not in a district.

The GMAs have the powers of a district.

If the powers of the GMAs are not curtailed, restrictive rules and regulations will be imposed within the GMA, by the GMA, without a vote by those living within its boundaries.

Today, statewide associations and organizations - and Texas property owners – are finally facing what Rural/Urban Resources first started warning about in 2006. These warnings were ignored by too many of the state's property owners.

It took four long years, to mid-2010, for the state's property owners to start a state-wide effort to try and stop the raid on the property owners' assets of groundwater.

20 – WATER MARKETERS

Today, government water planners quietly, persistently, talk under the table about "water thieves."

They have even developed a disparaging term they can use openly: Water Marketers.

It is usually said with a contemptuous smirk.

It is a term they use to describe a private group that offers water for sale.

It comes from a false knowledge that only government can do what needs to be done, which is a serious form of mental sickness or deviation.

Think about it. It is a term used by water planners and their supporters for owners and developers of private water resources. But water districts, which are governmental units, including underground water districts, develop, treat and sell water, are water marketers.

The dirty little secret-mudball is that legislators who believe that government, and who see themselves as a key part of it, needs to control all things, and those who owe their urban, big city supporters for support, have given the state's underground water districts the power to develop their own water fields and drill their own wells and to sell and transport underground water long distances.

Well, whoa, Nellie!

Its adjustment time.

How can the water planners and their equally mentally-unbalanced supporters not consider underground water districts to be water marketers? Or that water supply districts and cities, large and small, are not water marketers?

How much of a double-standard do we allow? Talk about hypocrisy . . . talk about stupidity.

Water scarcity and water quality are among the top environmental issues of the 21st century.

All living things need water to survive, as does the current world economy. As some wise man once said, "You can live without love, but not without water."

And yet, water is chronically under-valued.

The reason for the under valuation is two-fold:

First, large metropolitan centers act as if the supply was unlimited, and continue to focus on unrestrained growth, making low-cost deals with big water users.

Secondly, development of water resources has been a governmental endeavor, and the use of tax-dollars in construction of surface reservoirs and supply pipelines distorts the cost of water and, in turn, its market value.

The cost of water is relatively cheap to business and industry, because of price incentives given by cities to big water users: The more you use, the less it cost. In today's world, unlike energy or materials, the cost of water has little risk to a business' return on investment.

Why should a business think long-term about water-related risks when they face far more costly and immediate issues?

Why should government think about what it is doing wrong in the way it is doing water resource development? Especially, when government seldom ever considers that it does anything wrong.

Without a market value, the heavy hand of government is incapable of solving most long-term water shortage problems. Government is good at thinking long-term about what it wants, but seldom considers the cost – to humanity and in money – of what it wants to do.

Market value is not high on government's "water marketeering" agenda.

Both business and government face a big challenge in the localized nature of water problems. Some areas have water. Some areas don't. People and businesses should be free to move to where the water is, but it is far easier to ask government to move the water to them – especially, when someone else is paying for it.

Government can take the water in one area and give it to another area that needs it by force, but not at true cost. The cost of getting the water to the users is subsidized by the state's tax payers.

Market value never enters into the actual cost of water.

Unfortunately, businesses are largely unaware of how water-related risks will likely exacerbate the risks in many economic sectors in the near future.

The risks are difficult for companies to assess, due to poor information about the underlying supply conditions, unknown governmental actions in water resource development (and the potential reactions of their citizens to the actions), and fragmentary or inadequate reporting by cities and communities where a plentiful supply of water may or does exists.

Increasing water demand is an important challenge. This challenge has definite impacts on business, which include:
- ✓ Higher costs for water.
- ✓ Conflicts with local communities and other large-scale water users.
- ✓ Decreased amount of water available for business activities.
- ✓ Increased costs for water.
- ✓ Operational disruptions and associated financial loss.

Government Impacts on future growth and licenses to operate.
- ✓ Regulatory restrictions for specific industrial activities and investments.

✓ Increased responsibility (and costs) to implement community water infrastructure.

Taken together, this means that businesses will face vastly increased uncertainty about the availability and quality of their water supplies. It highlights the intensifying conflict between profits and water availability.

More and more companies are now starting to factor in water availability when deciding where to locate new facilities.

Instead of letting market values work by encouraging private water development endeavors, government continues to interrupt market forces.

They see private water resource development as something evil or unworthy.

In truth, without government interference, water resource development carried out by private water development firms is likely the only way that a true market value for water will be found.

How free are we as a society if we accept the view that a government endeavor is good and an equal or similar private sector endeavor is bad?

What rational discussion is possible when one side sets the terms of an argument, the definition of the words to be used in the argument, and reserves the right to determine the value of the terms and the definitions?

No one in a rational frame of mind would enter into a card game where the professionals on the other side set the rules and the value of the chips, and can change the rules and the values of the chips at their whim. . . But this is how the groundwater control game is being played in Texas today.

The term, water marketers, said with a contemptuous sneer, is reserved for a private endeavor of water development and sale. Yet, thousands of supposedly intelligent people accept the disparagement and nod their head in acceptance and agreement, without any consideration of the fact that all water sales are by both private and public water marketers.

If you were to ask the majority of big city voters if private water marketers should be prevented from developing and selling underground water, the likely answer would be a resounding, positive YES!

But cities, counties, municipal supply districts, water supply companies, including water control districts, are all in the business of developing and/or treating and/or selling water. They are all "water marketers".

The battle of public perception has been lost to the water planners.

It is also likely that those same big city voters would have little to no reservation about voting to give the state the right to take control of a landowner's underground property.

For years, weak-minded, morally-challenged politicians who do not understand the meaning of private property have told voters it is what needs to be done.

Because of the landowner's misplaced trust in their elected officials, or their apathy in looking after their own interest, they are now in danger of being rushed into courts before judges who also may not regard private property highly or who have been listening to the whispers and shouts of the professional water planners.

If landowners do not take the time or make the effort to shape or influence the terms – the meanings of the words used – of the debate, they'll lose.

And the professional, government "water marketer" wins.

21 – WHY DISTRICTS ARE WANTED
Or Following the Money . . .

As we have seen, the objective of a Master Plan for the development of Texas' water resources is the establishment of a water control district, of some kind, over every area of the state.

What citizens are being told is that the objective is *a need to conserve and protect* groundwater.

BALDERDASH! BALONEY! BULL DURHAM! And a lot of expletives deleted...

It is about money.

It is a way to gain power and control over others and, thereby, gain financially. Everything else is just decoration on a huge cake of manure.

All the water control districts in the world cannot create more water. Oh, they can treat wastewater and desalinate salt water to potable standards and build more lakes, and squeeze blood out of proverbial turnips, but none of these activities will create more water.

And they can't make it rain.

Yet.

Yes, rainfall in one area can be captured in a downstream watershed lake, and groundwater can be moved from one area to another, but this does not create additional *new* water.

Neither is government, through rules and regulations imposed on the water control districts, creating any *new* water when it takes the groundwater under one area and moves it to another area. It is merely playing God (while breaking several of His commandants).

Government is deciding and determining who it will help and, to a large degree, who will succeed and who will fail.

That is not the role of a fair and just government.

It is just government listening to the money talk.

For instance, water control districts have been given the power to impose more restrictive conditions and high fees on permit applications for private development or non-government approved development and transfer of water, regardless of type or location or use.

Ah, the money: The fees.

The exorbitant fees.

For instance, water control districts can claim over one-half (50%) of the money for water developed and transferred outside their boundary. This is for water the district doesn't own...just controls.

Yes, they can demand 50-cents (and more) of every dollar that should go to the owners, the property owners. Basically, they are claiming that half of the groundwater beneath the property owners' land belongs to the district.

They do this under currently existing contradictions in the Texas Water Code.

Legislators have incorporated this nonsense into the Texas Water Code. Yet, the same law-makers claim the property owner owns the groundwater. (It is stuff like this that makes millions of voters wonder what most legislators are doing in Austin.)

In Texas, government has to be the biggest program that property owners have ever seen that doesn't do anything for them.

Consider what legislators have done:

By law, (1) a fee-based district can demand a 50-percent export surcharge for water transferred out of the district; (2) a district supported by taxes can impose a rate the equivalent of the district's tax rate per hundred dollars valuation for each thousand gallons of water transferred out of the district or 2.5-cents per thousand gallons; or (3) a district may extort a "negotiated fee" between the district and the transporter.

These are fees that are in addition to a district's permit and production and other fees.

These are fees placed on the transfer of water from within the district to outside the district by private developers.

But these same fees do not apply to the district or another governmental unit that transfers water outside a district's boundaries. This is why a "negotiated fee" has been inserted into the Water Code.

If it isn't about the money, why would the water control districts not be free to set their own individual permit fee schedules?

But the water planners and their friends spelled out and put into writing and inserted in the Texas Water Code (§36.113), a fee schedule that far exceeds the costs that municipal wholesalers, such as cities, pay in Texas. Or what most large water users currently pay for the water they use.

It is more than just greed demanding exorbitant fees. It is theft on a grand scale, or legal extortion.

This is why millions of Texans view government today as much like an Egyptian pyramid with millions of bricks piled on top of each other, with no structural integrity, but done just by brute force and thousands of slaves.

Logic should tell us that the Texas water crisis is not an issue of scarcity, but lack of access in certain areas. Not even regions, but areas.

Highly populated metropolitan areas want additional water resources in order to keep pouring concrete, building high-rise apartments, shopping centers, and expanding for the sake of expansion.

And they're planning on doing all of it with some other area's water? Plus, large water users in private endeavors want water delivered to them while they are unwilling to go where the water is...? And all of them demand that the area where the water is located help them pay for taking it?

Fuzzy is as fuzzy does.

Folks, control is all about the money.

Every two-bit dictator, every "strongman" leader, the heads of centralized governments in countries large and small, those who take office intending to do good or bad, end up extremely rich and/or, sooner or later, dead.

To err is human, but to really foul things up, you need government. Those who believe that government is 100% right all of the time, seem to think that the only reason it doesn't work yet is because of those who keep spouting off about "rights" and a thing called "freedom".

Worldwide, just as in Texas today, water is front and foremost in the minds of those who believe strongly in a need for "social activism", based on their desire *to do good*.

They are so full of social awareness and consciousness that they sometimes forget what is the right thing they should be doing.

In this electronic age, an unholy merger of websites, blogs, videos, e-mails, social postings to Facebook, YouTube, and other sites, creates an unending stream of issues, theories, facts and lies around water issues. The internet has raised awareness and has created a digital, global think tank. It is very much about a philosophy that a lot of ordinary people can make an extraordinary difference; that they can successfully get "government to do something".

Mobilization around worldwide governmental control of water is gaining appeal, just as the public rallies around stamp out cancer or HIV/AIDS campaigns or an epidemic or disaster, such as what recently happened in Japan.

But like any movement, it is subject to misinformation provided by word masters who can paint a dire picture on the need for water. Supporters claim that plentiful, safe, clean water is a persistent problem that needs addressing. But, look at some of the organizations that are involved in water issues.

First, there is the United Nations, which recently declared that *clean water is a fundamental human right.*

In addition to that great world body of humanitarians, other organizations and groups include:
- Pan-American Health Organization
- UNICEF/WHO
- The World Bank Water Supply & Sanitation
- World Water Council
- International Development Research Centre
- Inter-American Water Resources Network
- International Institute for Sustainable Development
- International Office of Water
- World Health Organization

And these are only a drop in the water bucket. There are thousands of such associations, organizations, study groups, task forces, committees, etc., all arguing that clean water is a human right and they are working to make it a right.

They claim that a staggering number of people around the world, nearly a billion, are currently denied that right.

Some of the statistics that are thrown around are a little scary. For instance, it is claimed that every 20 seconds a child dies from water-related disease; that somewhere, each day, approximately 4,500 children die from unsafe water and lack of basic sanitation facilities; that a water and sanitation crisis claims more lives through disease than any war does through guns; that more people own cell phones than have access to a toilet; and the heart-rending statistics continue page after page.

You are told that in the time it took for you to read the above paragraph, at least one child died.

Should we be concerned for those facing such conditions?

Only a completely heartless entity, such as a government, which creates such conditions, would not be concerned. Hearts go out to innocent children and those who are oppressed.

But the reality is, as much as we are and should be concerned, we need to remember that those unfortunate people live under some form of centralized government. History shows that such governments are more concerned with rigid controls than helping those who live without the freedom of opportunity to improve their conditions.

If government was the answer, there would be no problems.

It is only a slight exaggeration to say that every government program represents a small failure of freedom.

Instead of government encouraging and enhancing opportunities for people to help themselves, government wants to do it. And when it does, not much gets done very quickly or as promised. But as they stand in the way of people getting it done in ways that are good for society, *they'll keep working to get it right.*

And millions – nay, billions – of dollars disappear into private accounts.

The same thinking driving the efforts of that great world humanitarian, the U.N., and many of the world's organizations and groups, is on display here at home, too. In addition to the blogs, websites, voluntary groups and other advocates, there are dozens and hundreds of nonprofit groups, government agencies and offices seeking more control of our nation's groundwater:

- The Peace Corp
- American Water Works Association
- Natural Resources Defense Council
- National Water Resources Association
- U.S. Environmental Protection Agency (EPA) Office of Ground Water and Drinking Water
- The U.S. Geological Survey, Water Resources of the United States

There are so many United States based nonprofit organizations committed to protecting and providing clean safe drinking water they are too numerous to even start listing; and they all "envision a day when everyone can have safe water."

There are dozens of such organizations in every state, including Texas.

But while all these organizations are involved in the spirit of the day, the popular current event or issue, they may be missing the point: While social good can become a powerful tool for change, it may not always be a change for the common good.

How far would you go?

Control of water equals control of people.

How far will you go?

Where does control out-weigh the freedom of private ownership? How can there be a "human" right without an underlying right of private property?

The "shakers and movers" – the power brokers – those who deal in power and control – have a direct relationship to a program and its ability to generate money for one or more of their friends, and thru their friends to them.

Whether it was "Papa Doc" in Haiti, Saddam Hussein in Iraq, Argentina's Perons, the Marcos in the Philippines, or Yemen, Kenya or some other "workers' paradise", a lot of the country's funds end up in private bank accounts hidden in some off-shore bank with high confidentially or non-disclosure standards.

And whether we like to admit it or not, many of our state and U.S. Senators and Representatives elected to multi-terms, end up with a lot more money coming out of office than they had going in . . .

The power to control equals money.

It is always about the money.

(A side note: The 2005 Corruption Perceptions Index listed over 150 countries of the world in their list of corrupt countries. Not one of the 150 countries is recognized as a bastion of freedom, but they all had one thing in common: A centralized government.)

Government is most always about the money.

In Texas, an acre-foot of water (325,681 gallons) is estimated to have about a $253 average cost. But in the Dallas-Fort Worth metroplex alone, another 96,000 acre-feet of water is currently needed annually.

That adds up to a lot of money.

Every year.

And when you consider the amount of water needed statewide even more money can be in play from water sales.

The water planners know that an immense amount of money is involved in the control of transferring water from one area in Texas to another. They *understand* that just half of the amount of money water could generate is worth all the efforts to control it.

You see, after all, when you follow the money, control is really just a way for those who will be in charge of the control to get to the money!

22 – WATER OWNERSHIP AND WATER USE

Depending upon which group is doing the espousing, Texas should be governed by a variety of water ownership or use rights – except for the one that has been established for over a century: Absolute Ownership.

Following a brief look at Absolute Ownership, are some of the "rights" that property owners have been un-burdened from or burdened with in various states, which water planners are urging for imposition on Texas' landowners:

Absolute Ownership: The earliest judicial theory of ground water rights is the doctrine of absolute ownership, also referred to as the English rule. Under the absolute ownership doctrine the landowner is, by virtue of land ownership, considered owner of the groundwater in place.

The English rule of absolute ownership reflected 19th-century judicial observations that the movement of groundwater was unknowable and thus it was unfair to hold a landowner liable for interfering with a neighbor's well when it was not knowable whether the defendant's pumping actually affected the plaintiff's well or not.

The English rule was once quite popular in the United States, and Texas is one of nine states that remain an absolute ownership jurisdiction; although Texas' legislators in recent years have passed a number of laws that have weakened the absolute ownership principle. Legislation has created a number of sub-state districts where groundwater use is now regulated, e.g. controlled, as in the Houston/Galveston area, High Plains, San Antonio.

The first groundwater district was formed in 1951, with headquarters in Lubbock, Texas. Today, all but 100-counties in Texas are either all or part in a control district.

Reasonable Use: The reasonable use rule, or American rule, was developed in the 19th century. Under the American rule landowners are entitled to use groundwater on their own land without waste. If their use exceeds this "reasonable use", the landowner is liable for damages. The American rule may still be followed in a few eastern states, although it is being judicially replaced by the eastern correlative rights doctrine. The reasonable use doctrine is part of the ground water jurisprudence of Nebraska, Arizona, and California.

Western Correlative Rights: The California doctrine of correlative rights was also initially developed in the 19th century but has continued to develop. Under the correlative rights doctrine, if the groundwater supply is inadequate to meet the needs of all users, each user can be judicially required to proportionally reduce use until the overdraft is ended. The policy significance of correlative rights is that each well owner is treated as having an equal right to groundwater regardless of when first use was initiated.

The correlative rights doctrine is part of the groundwater jurisprudence of California and Nebraska, although its sharing feature has been incorporated into the ground water depletion statutes of a few other western states.

Eastern Correlative Rights: The eastern correlative rights doctrine, inspired by the Second Restatement of Torts, states that when conflicts between users occur, water will be allocated to the "most beneficial" use, giving consideration to a wide variety of factors, including priority of use. Several factors are enumerated to be considered in a judicial determination of whether a water use at issue is "unreasonable": (1) the purpose of the interfering use, (2) the suitability of the interfering use to the watercourse, (3) the economic value of the interfering use, (4) the social value of the interfering use, (5) the extent and amount of harm it causes, (6) the practicality of avoiding the harm by adjusting the use or method of use of one riparian proprietor or the other, (7) the practicality of adjusting the quantity of water used by each proprietor, (8) the protection of existing values of water uses, land, investments, and enterprises, and (9) the justice of requiring the user causing the harm to bear the loss.

Statutory States

Permit States: A few states, including Florida, Iowa, Wisconsin, and Minnesota, require a state permit as a condition of well construction and use. Typically, users become subject to a rationing program during periods of shortage so that public water supplies are protected at the expense of other uses.

Appropriation States. With the exception of the major ground water using states (e.g., Texas, Nebraska, Arizona, and California), western states apply the doctrine of prior appropriation to groundwater. This means that the right itself is dependent upon obtaining a state permit rather than simply owning land overlying the groundwater supply. Between groundwater users, priority of appropriation gives the better right. This means that first in time is first in right.

Groundwater Rights

Groundwater rights conflicts between well owners, and problems caused by surface-groundwater interference are among the issues addressed by allocation law issues.

In the common law states, groundwater rights are based upon owning land overlying the groundwater supply and are defined by court decision. In the statutory states, including the eastern permit states, groundwater rights are based upon obtaining a state permit and complying with its terms. In the permit and appropriation states, state statutes generally define the extent of groundwater rights.

Common Law States: In all common law states, the right to use groundwater is based on owning land overlying the groundwater supply. In absolute ownership states, courts have ruled that malicious pumping or practices which may lead to contamination of the aquifer may be judicially restrained.

In reasonable use jurisdictions there is generally no ownership interest in the groundwater itself until it has been captured. Pumping may be judicially restrained to prevent waste or non-overlying uses.

In correlative rights jurisdictions the right to use groundwater is also based on owning land overlying the groundwater supply, although in California prescriptive rights can be obtained for non-overlying uses. Pumping may be judicially restrained to prevent waste or to apportion an inadequate supply. In eastern correlative rights states, pumping may be judicially restrained during shortages, although the basis upon which shortages will be allocated is not predictable. Groundwater rights are least well defined in the eastern correlative rights statutes, since judicial notions of what may constitute the "most beneficial" use of groundwater may change over time.

Statutory States: In both eastern permit states and appropriation states, rights to use groundwater are based on obtaining and complying with the terms of a state permit. However, most existing groundwater uses were automatically grandfathered into the permit system. Pumping rates may be limited in a permit and further limited during shortages. In eastern permit states, public water supply uses and domestic uses will generally be protected during shortages. In appropriation states, senior users (i.e., those with an earlier priority date, or in other words, an older well) are protected during shortages without regard to use. A junior user with a higher use, however, may be able to condemn a senior's use right during shortages and thus pump water out of priority.

Common Law States. In absolute ownership states, the landowner owns the water beneath the surface of his property. Thus, neither his neighbor nor the state can come on his property and take the groundwater. It is his to use or not use, to convey, assign, bequeath, reserve, lease, sell, etc.

The following pages are excerpts from the TEXAS WATER CODE, dealing with groundwater, and reaffirming many, if not most claims, arguments, instances, illustrations, or examples as presented.

The author does not claim nor state that the following contents from the Water Code are completely accurate or reliable or complete, as there is a "real time" lag from when the legislature acts and such actions are posted within the Water Code.

(*Italicized comments are those of the author.*)

The state's claim to control of groundwater first raised its head in "definitions" in Chapter 26 of the Texas Water Code. Note the first water defined is groundwater. When 'groundwater' was added to this section of the Code is unknown, or added by whom.):

CHAPTER 26. WATER QUALITY CONTROL
SUBCHAPTER A. ADMINISTRATIVE PROVISIONS
§ 26.001. DEFINITIONS.

As used in this chapter:

(5) "Water" or "water in the state" means groundwater, percolating or otherwise, lakes, bays, ponds, impounding reservoirs, springs, rivers, streams, creeks, estuaries, wetlands, marshes, inlets, canals, the Gulf of Mexico, inside the territorial limits of the state, and all other bodies of surface water, natural or artificial, inland or coastal, fresh or salt, navigable or nonnavigable, and including the beds and banks of all water-courses and bodies of surface water, that are wholly or partially inside or bordering the state or inside the jurisdiction of the state.

But, as proclaimed in the Texas Water Code, all surface water belongs to the state:

SUBTITLE B. WATER RIGHTS
CHAPTER 11. WATER RIGHTS
SUBCHAPTER A. GENERAL PROVISIONS
§ 11.021. STATE WATER. (a) The water of the ordinary flow, underflow, and tides of every flowing river, natural stream, and lake, and of every bay or arm of the Gulf of Mexico, and the storm water, floodwater, and rainwater of every river, natural stream, canyon, ravine, depression, and watershed in the state is the property of the state.

(b) Water imported from any source outside the boundaries of the state for use in the state and which is transported through the beds and banks of any navigable stream within the state or by utilizing any facilities owned or operated by the state is the property of the state.

And as recognized by the Texas Water Code, all groundwater belongs to the landowner: All the rights of ownership – except those rights limited or altered by rules promulgated by a district:

CHAPTER 36. GROUNDWATER CONSERVATION DISTRICTS
SUBCHAPTER A. GENERAL PROVISIONS
§ 36.002. OWNERSHIP OF GROUNDWATER. The ownership and rights of the owners of the land and their lessees and assigns in groundwater are hereby recognized, and nothing in this code shall be construed as depriving or divesting the owners or their lessees and assigns of the ownership or rights, except as those rights may be limited or altered by rules promulgated by a district. A rule promulgated by a district may not discriminate between owners of land that is irrigated for production and owners of land or their lessees and assigns whose land that was irrigated for production is enrolled or participating in a federal conservation program.

Under the Texas Water Code, Groundwater Districts can be created several ways:

(1) by majority vote of the citizens living within the boundaries of the proposed district;

(2) at the request of the local county commissioners;

(3) annexation of an adjacent county by an existing district; and

(4) a Priority District can be created, without a vote or a hearing, by state agencies. (A Priority District can even be created by a simple hearing, without an evidentiary hearing, over the objection of local property owners):

WATER CODE

CHAPTER 36. GROUNDWATER CONSERVATION DISTRICTS

SUBCHAPTER A. GENERAL PROVISIONS

§ 36.0151. CREATION OF DISTRICT FOR PRIORITY GROUNDWATER

MANAGEMENT AREA. (a) If the commission is required to create a district under Section 35.012(b), it shall, without an evidentiary hearing, issue an order creating the district and shall provide in its order that temporary directors be appointed under Section 36.016 and that an election be called by the temporary directors to authorize the district to assess taxes and to elect permanent directors.

(b) The commission shall notify the county commissioners' court of each county with territory in the district of the district's creation as soon as practicable after issuing the order creating the district.

Under the Texas Water Code, the following steps outline the process of carrying out the creation of a Priority District, including a tax levy. If the majority of the votes cast at the election are against the levy of a tax, a district can set permit fees to pay for the district's regulation of groundwater - including fees based on the amount of water to be withdrawn from a well. See below:)

WATER CODE
CHAPTER 36. GROUNDWATER CONSERVATION DISTRICTS
SUBCHAPTER A. GENERAL PROVISIONS
§ 36.0171. TAX AUTHORITY AND DIRECTORS' ELECTION FOR DISTRICT IN A PRIORITY GROUND-WATER MANAGEMENT AREA. (a) For a district created under Section 36.0151, not later than the 120th day after the date all temporary directors have been appointed and have qualified, the temporary directors shall meet and order an election to be held within the boundaries of the proposed district to authorize the district to assess taxes and to elect permanent directors.

(b) In the order calling the election, the temporary directors shall designate election precincts and polling places for the election. In designating the polling places, the temporary directors shall consider the needs of all voters for conveniently located polling places.

(c) The temporary directors shall publish notice of the election at least once in at least one newspaper with general circulation within the boundaries of the proposed district. The notice must be published before the 30th day preceding the date of the election.

(d) The ballot for the election must be printed to provide for voting for or against the proposition: "The levy of a maintenance tax by the _____ Groundwater Conservation District at a rate not to exceed _____ cents for each $100 of assessed valuation." The same ballot or another ballot must provide for the election of permanent directors, in accordance with Section 36.059.

(e) Immediately after the election, the presiding judge of each polling place shall deliver the returns of the election to the temporary board, and the board shall canvass the returns, declare the result, and turn over the operations of the district to the elected permanent directors. The board shall file a copy of the election result with the commission.

(f) If a majority of the votes cast at the election favor the levy of a maintenance tax, the temporary board shall declare the levy approved and shall enter the result in its minutes.

(g) If a majority of the votes cast at the election are against the levy of a maintenance tax, the temporary board shall declare the levy defeated and shall enter the result in its minutes.

(h) If the majority of the votes cast at the election are against the levy of a maintenance tax, the district shall set permit fees to pay for the district's regulation of groundwater in the district, including fees based on the amount of water to be withdrawn from a well.

According to the Texas Water Code, the operation of a district is an expensive endeavor, especially with the consideration of consultant fees: and compensation for directors and managers and staff:

WATER CODE

CHAPTER 36. GROUNDWATER CONSERVATION DISTRICTS

SUBCHAPTER A. GENERAL PROVISIONS

§ 36.057. MANAGEMENT OF DISTRICT. (a) The board shall be responsible for the management of all the affairs of the district. The district shall employ or contract with all persons, firms, partnerships, corporations, or other entities, public or private, deemed necessary by the board for the conduct of the affairs of the district, including, but not limited to, engineers, attorneys, financial advisors, operators, bookkeepers, tax assessors and collectors, auditors, and administrative staff.

(b) The board shall set the compensation and terms for consultants.

According to the Water Code, the operation of a district becomes even more expensive with the compensation for directors and managers and staff. The Code taketh away, and the Code giveth: By a simple resolution, a district's board can authorize payment of higher fees:

§ 36.060. FEES OF OFFICE; REIMBURSEMENT. (a) A director is entitled to receive fees of office of not more than $150 a day for each day the director actually spends performing the duties of a director. The fees of office may not exceed $9,000 a year.

(b) Each director is also entitled to receive reimbursement of actual expenses reasonably and necessarily incurred while engaging in activities on behalf of the district.

(c) In order to receive fees of office and to receive reimbursement for expenses, each director shall file with the district a verified statement showing the number of days actually spent in the service of the district and a general description of the duties performed for each day of service.

(d) Section 36.052(a) notwithstanding, Subsection (a) prevails over any other law in conflict with or inconsistent with that subsection, including a special law governing a specific district unless the special law prohibits the directors of that district from receiving a fee of office. If the application of this section results in an increase in the fees of office for any district, that district's fees of office shall not increase unless the district's board by resolution authorizes payment of the higher fees.

Miscellaneous information from the Texas Water Code:

Sec. 36.020. BOND AND TAX PROPOSAL. (a) At an election to create a district, the temporary directors may include a proposition for the issuance of bonds or notes, the levy of taxes to retire all or part of the bonds or notes, and the levy of a maintenance tax. The maintenance tax rate may not exceed 50 cents on each $100 of assessed valuation.

(b) The board shall include in any bond and tax proposition the maximum amount of bonds or notes to be issued and their maximum maturity date.

As stated, a district can do what it refuses to allow others to do without paying high or exorbitant fees to the district:

Sec. 36.104. PURCHASE, SALE, TRANSPORTATION, AND DISTRIBUTION OF WATER. A district may purchase, sell, transport, and distribute surface water or groundwater.

The State Water Plan by and the opinions of the Texas Water Development Board determines a district's activities:

Sec. 36.1071. MANAGEMENT PLAN. (a) Following notice and hearing, the district shall, in coordination with surface water management entities on a regional basis, develop a comprehensive management plan which addresses the following management goals, as applicable:

(4) consider the water supply needs and water management strategies included in the adopted state water plan.

Sec. 36.1072. TEXAS WATER DEVELOPMENT BOARD REVIEW AND APPROVAL OF MANAGEMENT PLAN. (a) A district shall, not later than three years after the creation of the district or, if the district required confirmation, after the election confirming the district's creation, submit the management plan required under Section 36.1071 to the executive administrator for review and approval.

Sec. 36.1073. AMENDMENT TO MANAGEMENT PLAN. Any amendment to the management plan shall be submitted to the executive administrator within 60 days following adoption of the amendment by the district's board. The executive administrator shall review and approve any amendment which substantially affects the management plan in accordance with the procedures established under Section 36.1072.

The permit requirements don't seem too tough now, but it is getting more difficult – and costly.

TEXAS WATER CODE
§ 36.113. PERMITS FOR WELLS; PERMIT AMENDMENTS.

(e) The district may *impose more restrictive permit conditions on new permit applications* and permit amendment applications to increase use by historic users if the limitations:

(1) apply to all subsequent new permit applications and permit amendment applications to increase use by historic users, regardless of type or location of use;

(2) bear a reasonable relationship to the existing district management plan; and

(3) are reasonably necessary to protect existing use.

This section is aimed primarily at private developers as governmental units work together. There are no provisions for the districts to charge themselves or even to pay the area's landowners. It is a nice racket. And you have to wonder how they see a "reasonable" fee?

TEXAS WATER CODE
Sec. 36.122. TRANSFER OF GROUNDWATER OUT OF DISTRICT.

(a) If an application for a permit or an amendment to a permit under Section 36.113 proposes the transfer of groundwater outside of a district's boundaries, the district may also consider the provisions of this section in determining whether to grant or deny the permit or permit amendment.

(b) A district may promulgate rules requiring a person to obtain a permit or an amendment to a permit under Section 36.113 from the district for the transfer of groundwater out of the district to:

(1) increase, on or after March 2, 1997, the amount of groundwater to be transferred under a continuing arrangement in effect before that date; or

(2) transfer groundwater out of the district on or after March 2, 1997, under a new arrangement.

(c) Except as provided in Section 36.113(e), the district may not impose more restrictive permit conditions on transporters than the district imposes on existing in-district users.

(d) The district may impose a reasonable fee for processing an application under this section. The fee may not exceed fees that the district imposes for processing other applications under Section 36.113. An application filed to comply with this section shall be considered and processed under the same procedures as other applications for permits under Section 36.113 and shall be combined with applications filed to obtain a permit for in-district water use under Section 36.113 from the same applicant.

(e) The district may impose a reasonable fee or surcharge for an export fee using one of the following methods:

(1) a fee negotiated between the district and the transporter;

(2) a rate not to exceed the equivalent of the district's tax rate per hundred dollars of valuation for each thousand gallons of water transferred out of the district or 2.5 cents per thousand gallons of water, if the district assesses a tax rate of less than 2.5 cents per hundred dollars of valuation; or

(3) for a fee-based district, a 50 percent export surcharge, in addition to the district's production fee, for water transferred out of the district.

(f) In reviewing a proposed transfer of groundwater out of the district, the district shall consider:

(1) the availability of water in the district and in the proposed receiving area during the period for which the water supply is requested;

(2) the projected effect of the proposed transfer on aquifer conditions, depletion, subsidence, or effects on existing permit holders or other groundwater users within the district; and

(3) the approved regional water plan and certified district management plan.

(g) The district may not deny a permit based on the fact that the applicant seeks to transfer groundwater outside of the district but may limit a permit issued under this section if conditions in Subsection (f) warrant the limitation, subject to Subsection (c).

(h) In addition to conditions provided by Section 36.1131, the permit shall specify:

(1) the amount of water that may be transferred out of the district; and

(2) the period for which the water may be transferred.

(i) The period specified by Subsection (h)(2) shall be:

(1) at least three years if construction of a conveyance system has not been initiated prior to the issuance of the permit; or

(2) at least 30 years if construction of a conveyance system has been initiated prior to the issuance of the permit.

(j) A term under Subsection (i)(1) shall automatically be extended to the terms agreed to under Subsection (i)(2) if construction of a conveyance system is begun before the expiration of the initial term.

(k) Notwithstanding the period specified in Subsections (i) and (j) during which water may be transferred under a permit, a district may periodically review the amount of water that may be transferred under the permit and may limit the amount if additional factors considered in Subsection (f) warrant the limitation, subject to Subsection (c). The review described by this subsection may take place not more frequently than the period provided for the review or renewal of regular permits issued by the district. In its determination of whether to renew a permit issued under this section, the district shall consider relevant and current data for the conservation of groundwater resources and shall consider the permit in the same manner it would consider any other permit in the district.

(l) A district is prohibited from using revenues obtained under Subsection (e) to prohibit the transfer of groundwater outside of a district. A district is not prohibited from using revenues obtained under Subsection (e) for paying expenses related to enforcement of this chapter or district rules.

(m) A district may not prohibit the export of groundwater if the purchase was in effect on or before June 1, 1997.

(n) This section applies only to a transfer of water that is permitted after September 1, 1997.

(o) A district shall adopt rules as necessary to implement this section but may not adopt rules expressly prohibiting the export of groundwater.

(p) Subsection (e) does not apply to a district that is collecting an export fee or surcharge on March 1, 2001.

(q) In applying this section, a district must be fair, impartial, and nondiscriminatory.

Meeting the requirements as outlined in the Water Code, getting a permit from a district may not always be easy, nor cheap:

Sec. 36.1131. ELEMENTS OF PERMIT. (a) A permit issued by the district to the applicant under Section 36.113 shall state the terms and provisions prescribed by the district.

(b) The permit may include:

(1) the name and address of the person to whom the permit is issued;

(2) the location of the well;

(3) the date the permit is to expire if no well is drilled;

(4) a statement of the purpose for which the well is to be used;

(5) a requirement that the water withdrawn under the permit be put to beneficial use at all times;

(6) the location of the use of the water from the well;

(7) a water well closure plan or a declaration that the applicant will comply with well plugging guidelines and report closure to the commission;

(8) the conditions and restrictions, if any, placed on the rate and amount of withdrawal;

(9) any conservation-oriented methods of drilling and operating prescribed by the district;

(10) a drought contingency plan prescribed by the district; and

(11) other terms and conditions as provided by Section 36.113.

And a look at permits, as permitted by the Texas Water Code:

Sec. 36.115. DRILLING OR ALTERING WELL WITHOUT PERMIT.

(a) No person, firm, or corporation may drill a well without first obtaining a permit from the district.

(b) No person, firm, or corporation may alter the size of a well or well pump such that it would bring that well under the jurisdiction of the district without first obtaining a permit from the district.

(c) No person, firm, or corporation may operate a well without first obtaining a permit from the district.

(d) A violation occurs on the first day the drilling, alteration, or operation begins and continues each day thereafter until the appropriate permits are approved.

And if your neighbor doesn't like you:

Sec. 36.119. ILLEGAL DRILLING AND OPERATION OF WELL; CITIZEN SUIT. (a) Drilling or operating a well or wells without a required permit or producing groundwater in violation of a district rule adopted under Section 36.116(a)(2) is declared to be illegal, wasteful per se, and a nuisance.

(b) Except as provided by this section, a landowner or other person who has a right to produce groundwater from land that is adjacent to the land on which a well or wells are drilled or operated without a required permit or permits or from which groundwater is produced in violation of a district rule adopted under Section 36.116(a)(2), or who owns or otherwise has a right to produce groundwater from land that lies within one-half mile of the well or wells, may sue the owner of the well or wells in a court of competent jurisdiction to restrain or enjoin the illegal drilling, operation, or both. The suit may be brought with or without the joinder of the district.

(c) Except as provided by this section, the aggrieved party may also sue the owner of the well or wells for damages for injuries suffered by reason of the illegal operation or production and for other relief to which the party may be entitled. In a suit for damages against the owner of the well or wells, the existence of a well or wells drilled without a required permit or the operation of a well or wells in violation of a district rule adopted under Section 36.116(a)(2) is prima facie evidence of illegal drainage.

(d) The suit may be brought in the county where the illegal well is located or in the county where all or part of the affected land is located.

(e) The remedies provided by this section are cumulative of other remedies available to the individual or the district.

It may be your property, but the district has complete access to it:

Sec. 36.123. RIGHT TO ENTER LAND. (a) The directors, engineers, attorneys, agents, operators, and employees of a district or water supply corporation may go on any land to inspect, make surveys, or perform tests to determine the condition, value, and usability of the property, with reference to the proposed location of works, improvements, plants, facilities, equipment, or appliances. The cost of restoration shall be borne by the district or the water supply corporation.

(b) District employees and agents are entitled to enter any public or private property within the boundaries of the district or adjacent to any reservoir or other property owned by the district at any reasonable time for the purpose of inspecting and investigating conditions relating to the quality of water in the state or the compliance with any rule, regulation, permit, or other order of the district. District employees or agents acting under this authority who enter private property shall observe the establishment's rules and regulations concerning safety, internal security, and fire protection and shall notify any occupant or management of their presence and shall exhibit proper credentials.

The state's claim to control of groundwater first raised its head in "definitions" in Chapter 26 of the Texas Water Code:

CHAPTER 26. WATER QUALITY CONTROL
SUBCHAPTER A. ADMINISTRATIVE PROVISIONS
§ 26.001. DEFINITIONS.
As used in this chapter:
(5) "Water" or "water in the state" means groundwater, percolating or otherwise, lakes, bays, ponds, impounding reservoirs, springs, rivers, streams, creeks, estuaries, wetlands, marshes, inlets, canals, the Gulf of Mexico, inside the territorial limits of the state, and all other bodies of surface water, natural or artificial, inland or coastal, fresh or salt, navigable or nonnavigable, and including the beds and banks of all water-courses and bodies of surface water, that are wholly or partially inside or bordering the state or inside the jurisdiction of the state.

The following pages outline the powers and duties of the Texas Groundwater Protection Committee and provide a list of its membership, as of January, 2011.

The authorizing legislation for the Texas Groundwater Protection Committee is found in Chapter 26, of the Water Code:

CHAPTER 26. WATER QUALITY CONTROL
SUBCHAPTER J. GROUNDWATER PROTECTION
§ 26.403. CREATION AND MEMBERSHIP OF TEXAS GROUNDWATER PROTECTION COMMITTEE. (a) The Texas Groundwater Protection Committee is created as an interagency committee to coordinate state agency actions for the protection of groundwater quality in this state.

§ 26.404. ADMINISTRATION. (a) The committee shall meet not less than once each calendar quarter at a time determined by the committee and at the call of the chairman.

(b) Each member of the committee serves on the committee as an additional duty of the member's office and is not entitled to compensation for service on the committee.

(c) Each member of the committee may receive reimbursement for actual and necessary expenses in carrying out committee responsibilities as provided by legislative appropriations. Each member who is a representative of a state agency shall be reimbursed from the money budgeted to the member's state agency.

(d) Each agency listed in Sections 26.403(c)(1) through (8) of this code that is represented on the committee shall provide staff as necessary to assist the committee in carrying out its responsibilities.

(e) The committee is subject to Chapter 2001, Government Code, Chapter 551, Government Code, and Chapter 552, Government Code.

CHAPTER 26. WATER QUALITY CONTROL
SUBCHAPTER J. GROUNDWATER PROTECTION
§ 26.405. POWERS AND DUTIES OF COMMITTEE. The committee shall, on a continuing basis:

(1) coordinate groundwater protection activities of the agencies represented on the committee;

(2) develop and update a comprehensive groundwater protection strategy for the state that provides guidelines for the prevention of contamination and for the conservation of groundwater and that provides for the coordination of the groundwater protection activities of the agencies represented on the committee;

(3) study and recommend to the legislature groundwater protection programs for each area in which groundwater is not protected by current regulation;

(4) file with the governor, lieutenant governor, and speaker of the house of representatives before the date that each regular legislative session convenes a report of the committee's activities during the two preceding years and any recommendations for legislation for groundwater protection; and

(5) publish the joint groundwater monitoring and contamination report required by Section 26.406(c) of this code.

Texas Groundwater Protection Committee Membership:

Chairman — Texas Commission on Environmental Quality
Mark Vickery, Executive Director, MC-109
Texas Commission on Environmental Quality
PO Box 13087
Austin TX 78711-3087
Telephone: 512-239-3900

Designated Chairman:
Cary Betz, Groundwater Technical Specialist
Water Supply Division, MC-154
Texas Commission on Environmental Quality
PO Box 13087
Austin TX 78711-3087
Telephone: 512-239-4506

Vice-Chairman — Texas Water Development Board
J. Kevin Ward, Executive Administrator
Texas Water Development Board
PO Box 13231
Austin TX 78711-3231
Telephone: 512-463-7850

Designated Vice-Chairman:
Bill Hutchinson, PhD, PG, PE, Director
Groundwater Resources Division
Texas Water Development Board
PO Box 13231
Austin TX 78711-3231
Telephone: 512-463-5067

Railroad Commission of Texas
John Tintera, Executive Director
Railroad Commission of Texas
PO Box 12967
Austin TX 78711-2967
Telephone: 512-463-7068
Fax: 512-463-7000

Designated Representative:
Leslie Savage, Assistant Director
Railroad Commission of Texas
PO Box 12967
Austin TX 78711-2967
Telephone: 512-463-7308
Fax: 512-463-7005

Texas State Soil and Water Conservation Board
Rex Isom, Executive Director
Texas State Soil and Water Conservation Board
PO Box 658
Temple TX 76503-0658
Telephone: 254-773-2250
Fax: 254-773-3311

Designated Representative:
Donna Long, Water Quality Specialist
Texas State Soil and Water Conservation Board
PO Box 658
Temple TX 76503-0658
Telephone: 254-773-2250, ext. 228
Fax: 254-773-3311
E-mail: dlong@tsswcb.state.tx.us

Texas Department of Agriculture
Todd Staples, Commissioner
Texas Department of Agriculture
PO Box 12847
Austin TX 78711-2847
Telephone: 512-463-1408
Fax: 800-831-3884

Designated Representative:
Ambrose Charles, PhD
Deputy Assistant Commissioner for Pesticides
Texas Department of Agriculture
PO Box 12847
Austin TX 78711-2847
Telephone: 512-463-7699
Fax: 888-216-9834

Department of State Health Services
David Lakey, MD, Commissioner
Department of State Health Services
1100 West 49th Street
Austin TX 78756
Telephone: 512-458-7375
Fax: 512-458-7477

Designated Representative:
Ken Ofunrein, Group Manager
Compliance Inspections Group South
Environmental & Consumer Safety Section
Department of State Health Services
1100 West 49th Street
Austin TX 78756
Telephone: 512-834-6770, ext. 2451
Fax: 512-834-6644

Texas Department of Licensing and Regulation
David Gunn
Texas Department of Licensing and Regulation
Well Driller/Pump Installer/Abandoned Well Referral Program
PO Box 12157
Austin TX 78711
Telephone: 512-463-7880
Fax: 512-463-8616

Designated Representative:
Same

Texas Alliance of Groundwater Districts
Jim Conkwright, Manager
High Plains UWCD #1
President, Texas Alliance of Groundwater Districts
2930 Avenue Q
Lubbock TX 79411-2499
Telephone: 806-762-0181
Fax: 806-762-1834

Designated Representative
David Van Dresar, General Manager
Fayette County GCD
255 Svoboda Lane, Room 115
La Grange TX 78945
Telephone: 979-968-3135
Fax: 979-968-3194

Texas AgriLife Research
Mark Hussey, PhD, Director
Texas AgriLife Research
113 Jack K. Williams Building
2142 TAMU
College Station TX 77843-2142
Telephone: 979-862-3746

Designated Representative
B.L. Harris, PhD, Acting Director
Texas Water Resource Institute
2118 TAMU
College Station TX 77843-2118
Telephone: 979-845-1851
Fax: 979-845-8554

Bureau of Economic Geology
Scott Tinker, PhD, Director
Bureau of Economic Geology
Jackson School of Geosciences
The University of Texas at Austin
University Station, Box X
Austin TX 78713-8924
Telephone: 512-471-1534
Fax: 512-471-0140

Designated Representative
Bridget Scanlon, PhD, Senior Research Scientist
Bureau of Economic Geology
Jackson School of Geosciences
The University of Texas at Austin
University Station, Box X
Austin TX 78713-8924
Telephone: 512-471-8241

23 – WATER PLANNERS VS OWNERSHIP

Water planners and their short-sighted supporters argue that the Rule of Capture is flawed, saying "the fact that your neighbor can pump the water from under your property is a good example."

They *hope* for a system where "no one feels compelled to pump their water since we might need it for many years to come". They claim taking the owners' property is in "the public good."

They don't even want a "vested" claim, let alone absolute ownership. They are happy to politically plunder by the use of law in fuzzy court rulings.

They never seem to consider that it has never been proven that the "rule of capture" has done any harm.

In fact, the 'rule of capture' is the principle of law that allows each of us, including the planners, to retain ownership of what we own. But it is this principle they view as evil, and urge that it be set aside in order to take another's property. Oh, how shortsighted . . .

Why do they not understand that if this principle of law is set aside, they have no protection against someone taking their property? Oh, how foolish. . .

And they harp about "saving the water since we might need it for many years to come."

Okay: First, "we might need it" also means that it *might not* be needed". Especially if areas that need water would construct surface lakes to capture runoff from rainfall, water which the state already owns. Secondly, we're *saving it for whom?* Okay. Who will they be forced to save it for...? These questions are just as legitimate as the argument, and just as realistic. This is not saying that common sense conservation practices are not needed or that quality and quantity are not important. They are, but these things are a long way from taking the owners' property *for the public good.*

Hiding behind the posture of being "excellent land and water stewards", water planners offer some far-out excuses for taking groundwater. For instance, some are:

1) Without a system of protecting our water, you and others might lose your livelihood.
2) Right now, protection can only be found in the districts.
3) Use of water for a municipality outweighs the importance of all other water needs.
4) Limiting groundwater withdrawal ensures the protection of springs and rivers. And,
5) How can you ever truly measure and enforce vested water rights under your property versus that of your neighbor?

These are excuses of no or limited merit. Emotions are a poor substitute for facts. And damage occurs as the result of doing something we shouldn't...

For instance:

1) Fear of job or economic loss is instilling fear itself. If loss of water causes job loss in one area, it stands to reason that it will create another job and stimulate the economy in the area *with water.* Where is it written that planners have the authority (and especially *the wisdom*) to determine where jobs are created, and which areas gain and which areas lose from their actions?
2) As over 90% of all court cases dealing with groundwater have been and are those of landowners *filing against* a

groundwater district, their definition of "protection" is certainly open to question, if not to an interruption as downright strange.

3) The importance of a municipality is determined by the number of votes in it, *not need.*
4) It is a stretch to equate withdrawal of water from an aquifer to a river going dry. In far West Texas, rivers are usually dry beds, only seeing water from rainfall runoff. But aquifers exist under most of those dry river beds. If the argument was completely accurate, *shouldn't it work both ways?* And
5) Forget the "vested" nonsense the state's water districts and the Supreme Court are examining: The "measure and enforce" argument of the planners is the best argument *for landowners' groundwater ownership.* It is the basis for all the historical ownership rulings. It is the *not knowing* where the water comes from and where it is going and how it got to where it is that *is one of several reason* for "the rule of capture": For instance, studies have shown that water in some of the NE Texas aquifers seems to start in Arkansas and Oklahoma. Using the argument of state planners and their supporters, if the property owner doesn't own the water, wouldn't Arkansas and Oklahoma own it?

Talk about lawsuits!

Of course, if your main concern is *doing away* with private groundwater ownership rights, any argument that will hopefully advance your cause is permissible.

Unfortunately.

Changing social perspectives from private property to public ownership, *state ownership,* is a dangerous risk. For the state to take private property, for whatever reason, even with Supreme Court blessings, is *plunder by law;* and is a step into socialism that should be condemned by all who believe that an individual's right to own property is the fundamental foundation of human rights.

Planners know that if society and civilization is to progress, it is the job of government to control it for the public benefit of the masses.

They hold to a belief that channels of communications, industrial capacity, productive technologies, and natural resources managed (controlled) by government for "the public good" will make us invincible.

To the planners and their adherents, it is all about the *potentiality* to do *good*. It is the same failed logic as the Road to Nowhere. It is driving blindly past the detour signs that warn "Danger Ahead."

It is a quality that too many of us admire – and are influenced by it. But it is an attitude with the good of humanity squeezed out of it –

Planners do not believe in sin. How can there be sin when everything is being done for "the public interest; a public benefit, a public good"? Sin is, obviously, an attribute to man and, therefore, not a fallacy of government.

And the planners and their adherents blithely ignore the government debacles by the millions. (Don't ask us to explain stupidity.)

Water, of course, is Resource Number One.

It is the world's most vital commodity.

No one ever steals anything that is worthless.

Government never takes anything that is worthless.

The winners will be decided in battles in the legislature and before the court.

The Texas Water War is only beginning.

ABOUT THE AUTHOR

Jake Street, an out-spoken advocate for private property rights, was the first employee of Water, Inc. (a tri-state organization for the development of supplemental water resources for West Texas, Eastern New Mexico, and Western Oklahoma.)

JHe has served as manager of chambers of commerce and as a consultant to units of government; local, state and national.

His record of proven, concrete results includes helping in securing new industries for at least 11-different communities in 4-states; three community improvement programs he started received national publicity; and Industrial Relations Programs he initiated received six awards, including "Grand" and "First Place" Awards.

In his early years, Street worked all phases of the radio broadcasting industry, and as a political campaign manager.

Made in the USA
Charleston, SC
14 May 2011